IET POWER AND ENERGY SERIES 48

Series Editors: Professor A.T. Johns
Professor D.F. Warne

Wood Pole Overhead Lines

Wood Pole Overhead Lines

Brian Wareing

The Institution of Engineering and Technology

Other volumes in this series:

Volume 1 **Power circuit breaker theory and design** C.H. Flurscheim (Editor)
Volume 4 **Industrial microwave heating** A.C. Metaxas and R.J. Meredith
Volume 7 **Insulators for high voltages** J.S.T. Looms
Volume 8 **Variable frequency AC motor drive systems** D. Finney
Volume 10 **SF6 switchgear** H.M. Ryan and G.R. Jones
Volume 11 **Conduction and induction heating** E.J. Davies
Volume 13 **Statistical techniques for high voltage engineering** W. Hauschild and W. Mosch
Volume 14 **Uninterruptable power supplies** J. Platts and J.D. St Aubyn (Editors)
Volume 15 **Digital protection for power systems** A.T. Johns and S.K. Salman
Volume 16 **Electricity economics and planning** T.W. Berrie
Volume 18 **Vacuum switchgear** A. Greenwood
Volume 19 **Electrical safety: a guide to causes and prevention of hazards** J. Maxwell Adams
Volume 21 **Electricity distribution network design, 2nd edition** E. Lakervi and E.J. Holmes
Volume 22 **Artificial intelligence techniques in power systems** K. Warwick, A.O. Ekwue and R. Aggarwal (Editors)
Volume 24 **Power system commissioning and maintenance practice** K. Harker
Volume 25 **Engineers' handbook of industrial microwave heating** R.J. Meredith
Volume 26 **Small electric motors** H. Moczala *et al.*
Volume 27 **AC-DC power system analysis** J. Arrill and B.C. Smith
Volume 29 **High voltage direct current transmission, 2nd edition** J. Arrillaga
Volume 30 **Flexible AC Transmission Systems (FACTS)** Y-H. Song (Editor)
Volume 31 **Embedded generation** N. Jenkins *et al.*
Volume 32 **High voltage engineering and testing, 2nd edition** H.M. Ryan (Editor)
Volume 33 **Overvoltage protection of low-voltage systems, revised edition** P. Hasse
Volume 34 **The lightning flash** V. Cooray
Volume 35 **Control techniques drives and controls handbook** W. Drury (Editor)
Volume 36 **Voltage quality in electrical power systems** J. Schlabbach *et al.*
Volume 37 **Electrical steels for rotating machines** P. Beckley
Volume 38 **The electric car: development and future of battery, hybrid and fuel-cell cars** M. Westbrook
Volume 39 **Power systems electromagnetic transients simulation** J. Arrillaga and N. Watson
Volume 40 **Advances in high voltage engineering** M. Haddad and D. Warne
Volume 41 **Electrical operation of electrostatic precipitators** K. Parker
Volume 43 **Thermal power plant simulation and control** D. Flynn
Volume 44 **Economic evaluation of projects in the electricity supply industry** H. Khatib
Volume 45 **Propulsion systems for hybrid vehicles** J. Miller
Volume 46 **Distribution switchgear** S. Stewart
Volume 47 **Protection of electricity distribution networks, 2nd edition** J. Gers and E. Holmes
Volume 48 **Wood pole overhead lines** B. Wareing
Volume 49 **Electric fuses, 3rd edition** A. Wright and G. Newbery
Volume 50 **Wind power integration: connection and system operational aspects** B. Fox *et al.*
Volume 51 **Short circuit currents** J. Schlabbach
Volume 52 **Nuclear power** J. Wood
Volume 53 **Condition assessment of high voltage insulation in power system equipment** R.E. James and Q. Su
Volume 55 **Local energy: distributed generation of heat and power** J. Wood
Volume 56 **Condition monitoring of rotating electrical machines** P. Tavner, L. Ran, J. Penman and H. Sedding
Volume 905 **Power system protection, 4 volumes**

Published by The Institution of Engineering and Technology, London, United Kingdom

First published 2005

The Institution of Engineering and Technology
Michael Faraday House
Six Hills Way, Stevenage
Herts, SG1 2AY, United Kingdom

www.theiet.org

British Library Cataloguing in Publication Data

Wareing, B.
Woodpole overhead lines. – (IET power and energy series no. 48)
1. Overhead electric lines 2. Electric lines – Poles and towers
I. Title II. Institution of Electrical Engineers
621.3'1922

ISBN (10 digit) 0 86341 356 0
ISBN (13 digit) 978-0-86341-356-8

Typeset in India by Newgen Imaging Systems (P) Ltd, Chennai

Contents

Preface

This book considers wood pole distribution overhead lines. It is restricted to networks up to 132 kV only, although the fundamentals of lightning protection can be applied to higher voltage tower line networks as well. The bias is towards wood pole lines at medium voltage (11 to 33 kV) although much of the book will apply also to wood pole lines at 66 and 132 kV. Most of the book will be applicable to wood pole lines world-wide but, of necessity, some parts specifically relate to UK practice.

A comprehensive book covering electrical design, pole-top equipment, substations, asset management etc., would require many volumes. The intention instead is to concentrate on the mechanical aspects of distribution wood pole lines, including live line working, environmental influences, climate change and international standards. Lightning represents the major source of faults on overhead lines and a section is included explaining both how lightning affects lines and the strategic principles of protection.

This book covers topics such as wayleave, statutory requirements, safety, profiling, traditional and probabilistic design, wood pole design, weather loads, bare and covered conductors, different types of overhead systems, conductor choice, construction, condition assessment, maintenance, refurbishment, upgrading, lightning protection, live line working and modelling. The book shows that the subject is complex and demonstrates what constitutes good quality engineering.

Material for this book was generated for the IEE Wood Pole Overhead Lines School (WOLD) course, the Manchester University MSc course in power engineering, which included information from EA Technology Ltd, and also includes information supplied by many experts within the industry. I therefore acknowledge the source of much of this book to the WOLD contributors.

I am also grateful for the permission granted by the University of Manchester and EA Technology Ltd to use material from my work with them. In particular:

Information in chapter 2 is based on presentations given by Craig Robertson (consultant).

The material in chapter 3 is based on a WOLD presentation by Nick Minns and information from Steve Horsman of Optimal Technology and Poletec.

Chapter 4 contains material supplied by Craig Robertson as part of presentations on overhead line standards as well as presentations by the author as part of the University of Manchester power engineering MSc course.

Chapter 6 includes material presented at a WOLD course by Gordon McArthur, with updates and additions by Bill Sayer of Balfour Beatty Power Networks Ltd.

The material in chapter 7 is based on a WOLD course presentation by Bill Sayer as amended for an overhead line forum for DNO (distribution network operator) engineers.

Chapter 8 is based on two papers presented by John Evans and the author at a WOLD school and updated by Bill Sayer.

Chapter 10 is based on a presentation which was initially given by Paul Blezard (United Utilities) at a WOLD school.

Chapter 11 uses material provided by Craig Robertson, David Horsman (United Utilities) and Steve Horsman.

Chapter 12 includes material supplied by Steve Horsman.

Chapter 13 represents the work of Dave Hughes of EA Technology Ltd.

Chapter 15 includes information taken from papers on wood poles that form part of a two-day course on overhead lines given regularly at EA Technology Ltd by the author, a course module of an MSc power distribution engineering course at Manchester University, also by the author, plus information from David Sinclair of SIWT Projects Ltd and Steve Horsman.

Chapter 17 is based on a paper by David Horsman on the procedures used by United Utilities and on a presentation by Tony Pierce (Pierce Associates) at a WOLD school.

I am also indebted to Bill Sayer who reviewed the final draft.

I hope you enjoy the book and will dip into it many times. A glossary is provided (chapter 20) to allow you to keep track of all the abbreviations.

Chapter 1
The need for overhead lines

1.1 Scope

In this chapter, the problem of supplying power economically is covered from the various aspects of choosing underground or overhead, the route, the line voltage, wayleave and the various relevant electricity regulations. Although looking initially at new lines, much of this also applies to refurbishment or change of use of lines. By the end of this chapter, the different functions served by undergrounding or overhead construction will be established. Sometimes, ease of establishing an underground network can mean it is just as easy for someone to dig it up. Wayleave and environmental problems are major negatives for overhead lines, but faults are generally easier to find and quicker to repair compared with underground cables. However, this chapter will not solve this perennial argument – it is not intended to – but it may help an enlightened choice to be made.

1.2 New supply lines

1.2.1 Why are they required?

New overhead lines may be required on a permanent or temporary basis. Temporary by-pass lines may be required when a section of original line is being re-built or re-furbished in order to maintain power supplies to the customer. Alternatively, new out-of-town shopping centres or a new industrial estate may require power to be supplied to what was a green field site. These are examples where local demand requires the extension of the current network.

The building of new, small gas-fired or hydroelectric power stations or, as is more common now, wind farms, however, may require lines to connect the power source to a grid supply point (GSP). So, in this case, a whole new network may be required.

An increase in the local population may mean that the present supply system may not be able to cope. The choice is then to either upgrade the existing line or build a new line with capacity for future expansion.

Finally, new housing estates, schools, playing fields etc. may now surround an existing line that was once in open countryside. For reasons of safety and the environment the line may need to be diverted around the new development. So there can be many reasons why new supply lines are required.

1.2.2 Alternatives to new construction

Where it is not necessary or desirable to build new lines the alternatives are:

- line strengthening (replacing old components with new, e.g. poles)
- refurbishment (replacing old components with improved versions, e.g. switchgear)
- upgrading (increasing capacity by increasing voltage or conductor size)
- re-conductoring (replacing the old conductor with more suitable, e.g. safer conductors)
- re-design (changing the line design to avoid conductor clashing or susceptibility to lightning etc.).

1.2.3 OHLs and underground cables

1.2.3.1 General

Many people simply do not like overhead lines (OHLs) and consider them dangerous or ugly. There are several situations where lines are undergrounded for convenience or for environmental, visual impact or safety reasons. However, undergrounding can be an expensive process in urban areas or bad ground conditions. In open farmland the cost and speed of undergrounding in certain areas of the UK may be very little different from an overhead line construction, especially at low voltage or 11 kV. However, undergrounding in a granite area may not be an attractive alternative.

On overhead lines the major sources of faults are lightning, wind (blown debris, clashing etc.), snow/ice storms, ageing equipment and trees (growing or falling). National statistics show that underground cables suffer less faults per km than do overhead lines, but that these faults are more difficult to find, more expensive to repair and put consumers off supply for a longer time than overhead line faults. For all cables the main sources of faults are ground disturbance due to subsidence, road repairs, farming and other utilities laying or repairing their equipment. For oil-filled cables, oil leakage leads to insulation failure. The overall security level of an underground supply thus depends on line location and the closeness of other types of supply and transportation (gas pipes, water, TV network and telephone cables, roads). In the UK, over 200 companies are allowed to dig and lay underground services.

Undergrounding is cheaper in flat open areas with good soil conditions and easy access. However, underground cable is considerably more expensive than OHL conductor and requires joining to the local OHL network. Such junctions require lightning protection. Underground cables also require underground joints, which are a major source of oil leakage. The cost of tunnelling under immovable objects (motorways etc.) may be high.

Underground cable is not popular with farmers when fields are to be ploughed for crops as there is always a risk of ploughing up the cable. Also, there may be an area where deep ploughing is not allowed and the farmer is restricted from full use of the land.

Accurate maps are required for undergrounding in urban areas so that electricity cables do not suffer from inadvertent third-party disturbance. Unmapped or poorly mapped cable lines can be damaged in building projects or other excavations. Overhead lines may be undergrounded near small or large airports, in new housing developments and at crossing points with motorways, railways or high-voltage transmission lines.

1.2.3.2 Electrical considerations

Table 1.1 shows the basic electrical characteristics of overhead (OHL) and underground cables (UGC) at equivalent current ratings. Line capacitances and inductances are determined by the geometrical arrangement of the conductors but, in general, the inductance of an OHL is about three times that of an UGC. The capacitance, however, is the other way round. The capacitance of an UGC can be 20 to 30 times that of an overhead line. So the characteristic impedance of an OHL is about five to ten times higher than that of an UGC. This leads to an effective higher natural load on an OHL, i.e. the resistive plus the inductive load.

Current rating

The maximum allowable conductor temperature determines the current rating of an overhead line. Leaving aside the dynamic current ratings now being applied to increase power transfer, a line can generally handle around 3 A/mm^2. UGC current capacity is limited by the insulation and heat dissipation through the surrounding soil. Lifetimes are generally 50–60 years for OHLs, but only 35–50 years for cables. Reactive capacitance current in cables increases with voltage and line length and limits the power transfer capability – especially at higher voltages. UGCs can be

Table 1.1 Typical electrical data for OHL and UGC

Voltage		11 kV		132 kV	
Item	Unit	OHL	UGC	OHL	UGC
Material		ACSR[1]	Al[1]	ACSR[1]	Cu[1]
Size	Ω/km	120/20	240	265/35	630
Resistance	nF/km	0.24	0.125	0.11	0.03
Capacitance	mm^2	10	290	9.5	180
Inductance	mH/km	1.14	0.31	1.21	0.41

[1] ACSR – aluminium conductor steel reinforced, Al – aluminium, Cu – copper (see chapter 8).

solid dielectric (e.g. cross-linked polyethylene, XLPE) or oil- or gas-filled paper insulated.

Electro-magnetic fields (EMFs)

Public concern over a perceived EMF health problem can cause wayleave problems for OHLs. In the UK, EMF levels are normally several orders of magnitude below the national standards set by the NRPB. When considering replacing OHLs by UGCs, an informed assessment of the economic, technical and environmental considerations needs to be made.

Uprating/upgrading

Uprating or upgrading is easier to achieve with overhead lines by increasing the voltage level or the conductor size. This is more difficult to achieve on an UGC network.

1.2.3.3 Capital cost comparisons for overhead lines and underground cables

As has been shown, the costs of electricity supply can be extremely high even before permission has been given to build the line. The cost of the actual construction itself will depend on many factors, such as ground condition, deviations to avoid wayleave (e.g. build along/under a highway), re-instatement costs, termination costs (connecting with other parts of the network) etc. However, a rough idea can be gained from Table 1.2.

Table 1.2 is only a very rough guide and does not include the costs of obtaining wayleave etc., which in some cases can be far higher than the purely mechanical construction costs. However, access, bad ground and frequent roads etc. can make undergrounding expensive. Alternatively, in good open ground conditions with few obstacles, undergrounding can be considerably cheaper than, for example, an 11 kV overhead line.

Capital costs are often highlighted in OHL and UGC comparisons. There is no doubt that ploughing (the process of ploughing a narrow trench and laying in the

Table 1.2 Cost of overhead lines versus underground cables

Circuit type	Cost × £1000/km		Approximate ratio UGC/OHL
	Underground	Overhead	
132 kV tower line double circuit	1600	200–300	8:1 to 5:1
132 kV wood pole line single circuit	800	50–80	16:1 to 10:1
33 kV wood pole line single circuit	100	30–40	3:1 to 2.5:1
11 kV wood pole line single circuit	20–40	20–35	1:1 to 2:1
Low voltage	15	15	1:1

underground cable as one action) in suitable ground can achieve low capital cost undergrounding at 11 kV. However, there are other lifetime costs that need to be considered. These include operating and maintenance costs and the cost of losses. Over a lifetime the cost of UGC power losses can be highly significant and even outweigh any capital savings. The capital costs per MVA fall substantially as the operating voltage is increased since the power transmission increases far more quickly than the structure costs. For UGCs, higher voltages require higher insulation levels and the ability to withstand higher temperatures. Overall, the costs per MVA of undergrounding only fall off quite slowly at higher voltages, and at transmission voltage levels UGCs can be 18 times more expensive than OHLs [1].

1.2.3.4 Summary

The principal non-capital costs of a line are operational (maintenance) costs and the cost of losses (electrical). Extra capital costs may be incurred to meet regulatory or local safety requirements. Tower lines can last longer than wood pole lines but sometimes the line is not required to last long – if a major re-build is planned for ten years' time then maybe only a short life extension is required. Power losses in underground cables generate unwanted heat, and the dielectric loss increases with the voltage squared. There are therefore capacity and temperature limits for both overhead lines and underground cables. Total lifetime costs include inspection and maintenance as well as actual repair and component replacement. The full life evaluation must therefore take account of the technical, economic and environmental aspects of the line as well as the regulatory restraints that are now becoming more common.

1.3 Routing of overhead lines

1.3.1 Wayleaves and the 1989 Electricity Act

All electricity supply lines, whether overhead or underground, require consent from landowners that is known as wayleave. This allows the constructor to install and maintain the line. However, the landowner can also tell the utility to remove its line at any time under most wayleave agreements. The exception to this is if a line follows a public highway. That is one reason why roadside verges are dug up so frequently – wayleave is not needed.

Lines also require 'Section 37 consent'. This is explained in more detail in chapters 2–4, but essentially Section 37 of the Electricity Act 1989 requires that an electric line shall not be installed or kept installed above ground except in accordance with permission from the secretary of state. The exceptions to Section 37 are when an overhead line supplies a single customer (e.g. a farmer cannot object to being supplied with electricity for his/her own use) or when the line is within the premises of and under the control of the person responsible for its installation. That means that a factory can have its own supply lines on its own premises without asking permission. This also means that underground cables not in a public highway require wayleave consent.

Even before any supply can be considered, the problems of wayleave and Section 37 consent will arise. At times, these costs alone can make the construction of any new line totally uneconomic.

The 1989 Electricity Act states under Schedule I that the utility should:

> have regard to the desirability of preserving natural beauty, of conserving flora, fauna and geological and topographical features of special interest and of protecting sites, buildings and objects of architectural, historic or archaeological interest and shall do whatever can reasonably be done to mitigate any effect which the proposals would have on the natural beauty of the countryside or on any such flora, fauna, features, sites buildings or objects.

Such areas may be designated as sites of special scientific interest (SSSI) or as listed areas (by English Heritage). The Act also gives the right to object – possibly in a public enquiry – to electricity supply lines. So the cost of a new line is not only calculated in terms of poles, conductors and cross-arms, but also in the time and effort to obtain wayleave, Section 37 consent and any changes to the route or design forced by any conditions attached to the wayleave.

It will be easier, and therefore almost certainly cheaper, to obtain wayleave and Section 37 consent by a sensitive choice of line route. The major areas of concern for the public are generally visual impact, health and safety effects and financial depreciation of their property.

1.3.2 Visual impact

Objections can be based on the actual route (too close, obscures view) and the line design (low-profile wood pole commonly preferred to towers). Landowners may request undergrounding. Alternatives such as routing close to forest edges or keeping well below the skyline can also help. So it may help to work with the local topography and not against it.

1.3.3 Health and safety

There are two main areas of public concern over health and safety aspects of overhead lines. Schools, playing fields etc. as well as leisure areas (rivers) are seen as potentially dangerous areas for overhead lines. These dangers are highlighted in chapter 2. The problems can often be resolved by using covered instead of bare conductors.

There may also be public concern about the health effects of living near overhead lines and there have been many studies on the possible effects of magnetic fields generated by overhead lines. The National Radiological Protection Board (NRPB) has published a summary of all relevant work up to the year 2001 (see website for up-to-date information).

1.3.3.1 National Radiological Protection Board report

The NRPB reported on the possibility of leukaemia in children due to magnetic fields associated with tower lines. The report summary is available on the website http://www.nrpb.com.

The report states that out of the annual UK birth rate of 700 000, there will be around 500 cases of leukaemia in children. Two of these 500 cases would be associated with exposures to magnetic fields of 0.4 microtesla (μT) or more. This implies a risk of one case in two years due to the proximity of overhead lines. However, the report states that at this low level there may be biases in the way in which the data have been collected and that there is no good evidence at the moment that exposure to EMFs is involved in the development of cancer, and in particular leukaemia. Nor do other human or animal epidemiological studies suggest that EMFs cause cancer. There is some evidence that high levels of EMF are associated with a small risk, but these levels are not experienced in the UK. Overall, the current evidence is not strong enough to justify a firm conclusion that EMFs cause leukaemia in children. The report recommends further research.

The press release by the NRPB states: 'The review of experimental studies by AGNIR (the group that produced the report) gives no clear support for a causal relationship between extremely low frequency (i.e. power frequency) EMFs and cancer.' The Radiological Protection Act (1970) defines an investigation level of 1600 μT for power frequency magnetic fields and 12 kV/m for electric fields. In almost all cases actual values are well below these levels.[1]

1.3.4 Financial depreciation/sterilisation

There are restrictions on land use in the immediate vicinity of overhead lines and underground cables. Property values can be reduced by five per cent or more due to these restrictions, and visual impact and hence compensation may need to be negotiated.

These days the least expensive option in building a new line may be to follow environmentally sympathetic guidelines in determining the route. Although possibly incurring more expense in the mechanical construction, the overall project cost may be lower if landowners, councils, environmental groups etc. can be persuaded that the utility is on their side in protecting the environment. Consider especially:

1 Scenic, SSSI (sites of special scientific interest) or historically important areas. These can include marshes where migrating or sea birds congregate, water meadows for flora, relatively recent but listed buildings, such as in old coal mine areas, or historically important archaeological sites.
2 Use of topography and trees to hide lines. Routing lines along valleys, avoiding skylines and going along the edges of forests can reduce the visual impact of lines. Incorporating poles into hedge lines is another means by which lines can be hidden.
3 The different approach to residential and industrial areas. Residential areas are generally not acceptable for overhead lines, and this can apply to areas liable to become residential over the forthcoming ten years. On the other hand, industrial

[1] All the above statements have been taken from the report and are not conclusions drawn by the author or the IEE from reading the report.

areas are often used to power lines being visible but the companies may prefer delivery at 11 kV rather than 415 V, or even at voltages up to 132 kV. In such cases, the customer is then responsible for their own sub-station.

1.4 Summary

This chapter has looked at why new distribution lines may be necessary, and also at the alternatives of upgrading or extending the life of current lines. The commonly quoted alternative of undergrounding has also been addressed, albeit briefly. This is a precursor to a more detailed examination of this aspect of line construction and alteration in the next chapter. Environmental as well as safety aspects of overhead lines have also been touched on. Safety will become a more important factor to consider over the next few years as the new Electricity Supply Regulations (2002) are implemented. For this reason, the topic is also covered in more detail in chapter 4.

1.5 Reference

1 Cigré Session Paper 21/22-01, Paris, August 1996

1.6 Further reading

National Radiological Protection Board: 'ELF electromagnetic fields and the risk of cancer'. Report of the Advisory Group on Non-ionising radiation, 2001, **12**(1), pp. 173–179

For those interested in Distribution Overhead Lines in depth, some interesting reading is available. One excellent publication is the Southwire 'Overhead Conductor Manual' (first edition 1994). Although it has been out of print it is currently mentioned on the Southwire web-site

Another excellent and free reference which is definitely available is 'Design Manual for High Voltage Transmission Lines'. It is published by United States Department of Agriculture – Rural Electrification Administration and considers pole line design up to 230 kV. It has just been revised (September 2004) and can be downloaded from http://www.usda.gov/rus/electric/pubs/1724e-200.pdf

The document runs to over 300 pages and is a 7.6 MB Acrobat download. This site also refers to a raft of other interesting publications which are worthy of closer inspection. The electrical publications index can be found at http://www.usda.gov/rus/electric/bulletins.htm

Chapter 2
Statutory requirements

2.1 Introduction

The routing and construction of overhead lines is governed by a number of statutes and regulations. The first part of this chapter gives a broad outline of the relevant legislation relating to rights to erect a line and general safety and environmental considerations. The second part lists the key points of the main acts and regulations.

The four key acts of parliament and regulations that apply to overhead lines will be covered in detail, as they will be referred to in other parts of this book. This chapter aims to outline the statutory process required to obtain consent for overhead line changes and construction, and identify safety issues relating to overhead lines. In addition, the statutory obligations that have to be followed in the construction and maintenance of overhead lines are described in relation to:

- wayleave
- safety
- mitigation of environmental effects.

2.2 Planning and routing: key acts and regulations

2.2.1 Acts, regulations and standards

Assuming that a line is needed between point A and point B, or maybe there are to be significant changes to an existing line, no one can carry out the work unless they are a licensed electricity supply company or a sub-contractor of such a company. Even then there is a maze of statutory acts and electricity supply regulations that are there to ensure that basic minimum standards are adhered to. These are important to maintain consistent and safe practices across the UK. Novel types of overhead line can be constructed but only when a design is produced that shows that all the regulatory requirements have been met.

From the advent of the first electricity supplies in the 1880s, the development of standards has followed the same three-step pattern:

1 The government of the day enacts legislation to cover the functions of electricity supply – the electricity supply acts.
2 The appropriate government minister of the day, by virtue of the powers conferred upon him by the act, produces technical requirements in the form of a statutory instrument – the electricity supply regulations.
3 The generators and distributors of the day interpret the requirements of the acts and regulations into engineering standards.

Stage 1 and the legal side of stage 2 are covered in this chapter, and the technical part of stage 2 and the whole of stage 3 are covered in sections 4.1 and 4.2.

2.2.2 Planning consents and approvals

Section 37 of the 1989 Electricity Act states: 'An electric line shall not be installed or kept installed above ground except in accordance with a consent granted by the Secretary of State.'

There are two main exceptions to this rule:

- where a line does not exceed a voltage of 20 kV and will supply only one consumer (a service line)
- where the electric line will be built on land within the direct control of the utility.

The first exception was intended to cater for, for example, an individual farm on a long HV spur. Overhead line engineers will come up against Section 37 regularly – especially when sufficient modifications are made to a line to render it necessary for Section 37 approval. Over the last few years, this process has become less onerous as the responsibility for ensuring that a line is 'fit for purpose' has come to rest more with the distribution network operator (DNO) than in preprivatisation days.

2.3 The impact of the 1989 Electricity Act

Very significant changes have been made to the way the electricity supply industry (ESI) operates in the UK as a direct result of the 1989 Electricity Act. This Act still maintains that a properly engineered design is used and all construction, maintenance and operation is in a safe and efficient mode, but changes the onus of responsibility onto the supply company. The company has to ensure that within current standards the line is fit for the purpose for which it is intended. A line may be designed and built to a high enough standard in Kent, but it may not be suitable to provide a safe and secure supply in Scottish blizzards – so it would not be 'fit for purpose' in the north of Scotland.

Although only a few of these changes have had any immediate impact on wayleave departments in the wood pole world, obtaining and retaining consents for overhead lines is becoming increasingly difficult. Landowners and councils are now generally more aware of their legal position in respect to the ESI under the 1989 Act. This ensures that they can influence line routes and designs that, initially, were selected by the electricity company alone. They also realise that the industry is now a commercial enterprise. Therefore, in some cases it becomes almost impossible to obtain consents for new routes. Although it is possible to seek compulsory powers under the Act, in practice this is only very rarely applied to overhead lines below 132 kV. The costs of obtaining consent can be higher than those of the line construction.

As the voltage reduces, so does the cost differential between overhead and underground (see chapter 1), particularly at 11 kV. By the time statutory proceedings are instigated and a public enquiry is set up, the difference in cost may become negligible. Assuming wayleaves have been secured for the new line, there is then local authority and ministerial consent under Section 37 of the Act (see section 1.3.1) to be obtained. This of course has always been the case and is often a straightforward process, particularly with re-builds of existing lines. There are cases, however, where a local authority has a defined policy against overhead lines, and without its support the minister will not automatically give consent. Legally, the minister has the power to overrule it but in practice rarely will. The process is:

1 a new line is needed or a current line needs upgrading
2 a suitable design is obtained and a route specified
3 wayleave then needs to be obtained from the landowners along the route
4 the local authority and environmental groups must also be consulted
5 Section 37 consent must be obtained
6 if there are serious objections from landowners or others (e.g. English Heritage, Royal Society for Protection of Birds etc.) with an interest in the route then a public enquiry may be needed.

2.4 Wayleaves, easements and other methods of securing rights

It can be overwhelming at first to consider all the aspects associated with the introduction of an overhead line. These are initially:

- the electrical load requirements of the proposed overhead line
- the mechanical strength capabilities of the line components (conductor, fittings, steelwork and structures) used to meet with the environmental loads (wind, ice and snow)
- the components' reliability and longevity in the field and the techniques or methodologies adopted for the construction and maintenance of the overhead line.

As an overhead line design engineer, each of these issues must be addressed. However, a great deal of effort may reap little reward if a route cannot be agreed with local authorities and landowners. It is therefore necessary to provide flexibility in design to provide options for the routing of new lines. The following rights of landowners and occupiers all need to be considered:

- wayleaves
- necessary (compulsory) wayleave
- easements
- prescriptive rights (squatters' rights)
- adverse possession.

2.4.1 Wayleaves

Wayleaves have been in existence in one form or another for hundreds of years. It is simply an agreement between two parties whereby party B wishes to carry out some simple activity on party A's land. The two come to an agreement about what the activity is, how long it should go on for, any payment to be made by B to A in respect of the use of his land and any other matter the two feel it is in their mutual interest to agree upon. Of course, things can get much more complicated, but this, in its simplest form, is the basis of the wayleave consent that is granted to the DNOs for most of their overhead plant. Each DNO has its own particular wayleave consent form, which reflects the things they consider important, and, by standardising the agreed matters on a form, a quick and easy way to secure a right to place equipment on someone's land is provided.

A typical example of a wayleave consent form currently in use for voltages below 132 kV can be broken down into its constituent parts:

- It initially identifies the grantor as the person who is legally entitled to give consent, i.e. the owner or his/her authorised agent, and then states that he or she gives their consent to what follows later in the form.
- The first schedule describes exactly where the property in question is.
- The second schedule describes exactly what the DNO wants to do.
- The third schedule describes in detail the DNOs commitment to the grantor in terms of compensation for damage etc., and sets out such matters as annual payments of rent and compensation.
- Lastly, follows the signature of the grantor, his/her full address and a space for a witness to sign.

2.4.1.1 Payments for wayleave

A small rental payment is often made but this serves only to consolidate the legal position whereby the electricity undertaker could be forced to remove its plant from the grantor's property if it defaulted on payment. The sum of £4.00 per annum is currently the national norm for a single wood pole. Eventually, complaints from farmers caused a second amount to be paid – a compensation figure which varies with the type of land cultivation. Currently, this is £11.32 for a single wood pole in arable land.

2.4.1.2 Termination of consent

The wayleave consent still gives the landowner the right to give a DNO notice that he is withdrawing his consent and the DNO must take action to remove its equipment covered by the consent after a 12-month expiry period. In practice, the 12 months can be used by the DNO to investigate ways of removing or re-siting its equipment, or by re-negotiating its position with the grantor. Following this 12-month period, the DNOs equipment must be removed. If it is not, and no compromise is reached with the grantor, the DNO is in breach of the Act and can be pursued through the courts.

Any termination notice must be taken very seriously. In these situations, the DNO can appeal directly to the secretary of state for a 'necessary wayleave' under paragraph 8 of the fourth schedule of the Act. This effectively re-starts the 12 months' notice.

2.4.2 Necessary wayleave

Schedule 4 of the 1989 Act details what is often known as a compulsory wayleave. This gives a DNO the opportunity to appeal to the secretary of state for a wayleave that would not have been granted voluntarily by a landowner. It must be borne in mind that costs will be high as a public enquiry will almost certainly be needed and the outcome can be far from certain. This approach is usually reserved for the higher voltages where costs of alternatives outweigh the cost of the procedure.

2.4.2.1 Change of landowner

One of the biggest drawbacks with wayleave consent is the fact that it only binds the person who signs it – it does not bind the land. This is made clear in paragraph 8(c) of the fourth schedule of the Act. The practical consequence of this is that the wayleave is only legally valid for as long as the person who granted it owns the land. If the letter of the law were to be followed, a new wayleave would need to be sought each time the land on which equipment is placed changes hands. If a new owner refuses to accept money in respect of an existing consent and requires equipment to be removed, legally the DNO is trespassing and must remove its equipment after the expiry of a three-month notice period from the landowner. This all makes the use of roadside verges very attractive to DNOs as access rights and wayleaves are not required.

2.4.3 Easements

A much more secure way of obtaining a right to place equipment on private land is to secure an easement. An easement is an absolute right in law to carry out an activity on someone else's land. This gives by far the greatest protection to any equipment placed on private land and is extensively used by other statutory undertakers to secure their water and gas pipelines etc. Its two greatest benefits as far as the ESI is concerned are:

1. Unlike a wayleave, it cannot be rescinded by any person other than the beneficiary unless there is a clause included in the documentation to that effect.

2 It binds the land rather than the person who granted it, so again assuming there is no wording included to the contrary, it is attached to the title of the land and transfers with it. Consequently, whoever owns the land in the future is bound by it.

2.4.3.1 Additional protection

An easement can also be drawn up to include other things that could be relevant – for example a 'not to build' clause is usual as a way of protecting an underground cable from being built over. Similarly, a route can be reserved where a line passes over a potential gravel or other mineral-rich area. In this particular situation, it may not be in the DNOs best interest to remain stoically fixed on its route – it may be better to divert the line, but only after the person or company who is gaining the benefit of mineral extraction has agreed to meet the cost of such a diversion. In this way, a DNO can be protected from unexpected costs incurred by demands for the re-siting of its equipment. However, it is usually much more difficult to persuade a landowner to enter into an easement (or deed of grant) because it binds his land absolutely and it may become an encumbrance in the future. It can devalue potential development land, particularly if the easement carries a 'not to build' clause.

2.4.3.2 Costs

The easement can be very expensive depending upon individual circumstances. The effect an easement will have upon the land must be borne in mind – particularly if that land has development potential. Although each DNO will have guidelines to assist with the drawing up of easements, costs can easily run into many thousands of pounds, particularly on the longer runs of 33 kV or 132 kV cable. This must be balanced against the cost of moving the cable should a wayleave be terminated. This will usually far outweigh the cost of the easement.

2.4.4 Prescriptive rights (squatters' rights)

From time to time a DNO is placed in the situation whereby it is required to remove its equipment from private land and no wayleave or other agreement can be traced. This is particularly the case with very old low-voltage distribution systems with poles or cables in gardens. No consents were secured at the time the line was built. This happened for various reasons – the two most common seem to be:

1 It was taken for granted that if a supply were required there would be no objection to a pole on your land.
2 Often, it was the town council that operated the electricity supply undertaking and the houses to be supplied were owned by the council. This is particularly true in the case of mural wiring, where distribution wiring runs along the front wall of a row of terraced houses. However, if it can be proved that an item of plant has been *in situ* for a long time – usually 20 years – then there is a chance that a prescriptive right can be claimed. The DNO must be able to demonstrate that there has never been a request to remove the equipment in the last 20 years and the equipment was not placed or kept there in secret. It is difficult to imagine

how an overhead line could possibly have been kept 'in secret', but it is a much more complex matter to say the same about an underground cable. The problem is further compounded if the land on which the equipment stands has changed hands over the years. The prescriptive right can then be challenged as the period of time is generally considered to apply to one owner only.

2.4.5 Adverse possession

Adverse possession applies again in situations where a DNO has no agreement to carry out its activities on private land, but has been doing so for many years. The major difference here though is that, through its activities on the land, the DNO can claim the land as its own and carry on its occupation, e.g. at sub-stations or in other situations where a parcel of land is occupied. This can apply to large or small areas of land.

2.4.6 Other matters

Schedule 8 of the 1989 Act refers to other matters that need attention as part of an application under Section 37. Paragraph 1 states that a plan must include showing the land that the line will cross, the route length and voltage, and it must be made clear if all wayleaves have been obtained.

2.4.6.1 Local planning authorities

Under Schedule 8, a DNO is required to notify the local authority (LA) (both local and county at 132 kV) that an application is being made to the minister under Section 37. In practice, the LA will be consulted first and will probably consult also the relevant parish councils affected by the proposal. The difficulties start if an LA refuses the application. In this situation, the DNO must decide if it is willing to negotiate with the LA to achieve some compromise, withdraw its application completely or allow the secretary of state to set up a public enquiry.

Again, a DNO can avoid to some degree placing itself in this position by opening discussions with the LA at an early stage in the proceedings – even before wayleaves have been obtained. This will give a feel of how the LA views this particular application and the route can perhaps be modified to take account of its views.

The public enquiry itself is presided over by an inspector appointed by the minister, who will hear evidence from both sides and consider written information supporting the appellant (the one appealing for permission) or the objector. He or she will then close the enquiry when it is felt that all sides have made their cases and go away to consider the case in private. All this is expensive and time consuming.

2.4.6.2 The Electricity Supply Regulations 1988

This statutory instrument deals with the general operation of a distribution system and supersedes many individual instruments that were in force before. As far as overhead lines are concerned, Part IV defines the following conditions:

- where they can be placed (avoidance of danger)
- the minimum height above ground and roads
- insulation and protection
- precautions against access
- stay insulators.

Looking briefly at each item:

Where they can be placed. All overhead lines (except in a sub-station or generating station) must comply with these regulations.

Minimum clearances above ground and roads. These are defined in Schedule 2 of the document. The regulations specify the minimum height of overhead lines, wires and cables above ground at its likely maximum temperature – not ambient temperature – so maximum electrical load must be taken into account at the design stage. It also requires that 'All supplier's works shall be sufficient for the purposes for, and the circumstances in, which they are used and so constructed, installed, protected (both electrically and mechanically), used and maintained as to prevent danger or interruption of supply so far as is reasonably practicable'. The important part here is the 'sufficient for purpose'. This means the onus is on the supplier to make sure everything is safe. The clearances are shown in Table 2.1.

In practice, taking into account the height of today's agricultural machinery, which can exceed 4 m, to design a line to 5.2 m across fields likely to be used by such machinery is not to be recommended. In most DNOs all new lines are designed to a ground clearance of not less than 6.5 m at a conductor temperature of 65 °C. This gives a degree of tolerance, should it be needed at construction, and ensures that combine harvesters and sprayers etc. have plenty of clearance and may be operated without undue concern. However, even though the farmer may know very well that the line is there, there are many examples of injury or death due to accidental contact of farm vehicles with OHLs.

Table 2.1 Clearance requirements for OHLs

Voltage (kV)	Road (m)	Fields (m)
Up to 33	5.8	5.2
33–66	6.0	6.0
66–132	6.7	6.7
132–275	7.0	7.0
275–400	7.3	7.3

Insulation and protection. This section of the regulations starts by defining what is 'ordinarily accessible' and then goes on to describe what may and may not be placed in such an area. It also describes how plant must be insulated or earthed to prevent damage or danger and deals with insulation where a building is to be placed near a line, which may cause the line to become accessible.

Precautions against access. This states that steps must be taken to prevent unauthorised persons gaining sufficient access to place themselves in danger (anticlimbing guards on poles with attached plant), and high-voltage lines must carry 'danger' signs.

Stay insulators. All stay wires attached to any support carrying a bare live electric line must be fitted with an insulator, placed no lower than 3 m above the ground.

2.5 Statutory requirements – safety legislation

2.5.1 General

There are many ways in which people can injure themselves on electricity supply lines, and this is actually of more than just passing interest. The utility has to bear in mind everyone's safety when designing, constructing and repairing lines, and so the general pattern of public behaviour is of prime importance in determining risk levels. It is not always necessary for direct contact with a line to cause injury or death. Flashovers can occur to conducting objects such as fishing rods or metal poles or ladders if they approach close to a bare wire overhead line.

It is not within the scope of this book to cover operational risks to utility personnel, but it is essential and required by the Health and Safety Executive (HSE) that operational procedures are written to avoid risk or injury to linesmen or other utility personnel.

The safety of the public and of supply industry personnel or contractors is paramount. If the line is overhead someone may fly into it or drive a car into a support structure at some time. If it is underground someone may dig it up. Many people are killed or injured every year by accident or through vandalism or theft, even when the lines are intact and in good condition. More problems are caused when gales, electrical storms or blizzards bring lines down, eliminating the normal safety clearances between the high voltage and people going about their normal routines.

2.5.2 Areas where there is injury risk

Safety is always a delicate area to consider. The Health and Safety Executive (HSE) monitors safety aspects of the electricity supply network and is involved with all fatal enquiries. Table 2.2 gives a list of the fatalities in the agricultural industry caused by contact with an overhead line.

The following data and case studies are extracted from the HSE publication *Fatal accidents in the farming, fishing and forestry industry 1999–2000.* The following case studies describe the circumstances surrounding the deaths of the two individuals killed in 1998/9.

Table 2.2 Fatal injuries caused by contact with OHLs to employees and self-employed people in agriculture, 1986/7–1998/9 (extract)

Type of accident	Employees	Self-employed	Total
Hand tools	4	2	6
Overhead lines	23	8	31
Industrial plant	4	2	6
Domestic type equipment	0	3	3
Other	5	1	6
Total contact with electricity overhead lines	36	16	52

Case study 1: self-employed contractor 47 years old

A 47-year-old self-employed contractor was electrocuted when his aluminium ladder contacted a link conductor on 11 kV overhead power lines. He was trimming Leylandii trees close to a transformer pole in a private garden. The ladders were almost six metres long and contacted the lines as he attempted to change position.

Case study 2: employee 70 years old

A man was electrocuted when a single section ladder made contact with some 33 kV overhead power lines. The 70-year-old employee was working in an orchard picking apples and he decided to use the seven-metre long aluminium ladder rather than one of the shorter ones available, which would normally have been used. As he walked under the power lines the ladder either contacted one of the conductors or caused a flashover and he was fatally electrocuted.

There are also many areas where overhead lines can cause injury or death to people and wildlife. Obviously, when lines are damaged by road accidents or heavy snowfall etc., there will be danger to life. However, there are many cases where people can injure themselves by accidental contact even when the line is intact and in normal operation. These two aspects are covered in the following two sections. Agricultural examples are not included, as they have been considered in a previous exercise.

2.5.2.1 Fatalities when lines are intact

Forests

Timber trucks load timber by hydraulic cranes in the forests. Despite all safety instructions the crane can touch the overhead lines with potential injury risk to the operator.

Tree branches can also touch line conductors causing an earth fault but with too low a current to bring out the line protection. However, this fault current passing through the tree can pose a risk of electric shock to people or animals touching the tree. The Health and Safety Executive (HSE) specifies a 'safe' level of current passing through the human body of only 10 mA. For animals such as cows the dangerous level of current is around 1 mA (IEC 479-1).

Civil works

Overhead lines are often present near to or within building sites, and material transport vehicles and cranes can make accidental contact with these lines. Also, long conductive objects such as ladders or metal pipes (or even wet plastic pipes) may be carried by site workers and accidentally touch an overhead line. If the conductor is bare this type of incident will almost certainly have severe consequences.

Water crossings

Most of today's sailboats have plastic hulls with metal masts and stays. Although there are often clearly marked lanes with good clearance to overhead lines, when boats move in for 'overnighting' or travel outside these lanes due to human error the consequences of contact with a bare overhead line will probably be fatal to all on board. It is also important to recognise that people may sail on rivers not specifically intended for sailing. There is a major risk of electrocution from accidental contact between the mast and lines, especially when long spans are involved.

Fishing areas

Carbon fibre fishing poles are very popular. They have excellent rigidity and allow fishing to be done at a long distance from the fisherman. The sheer length of the poles and the attention of fishermen to their task mean that accidental contact with riverside overhead lines is always possible. This almost always leads to severe injury and is often fatal. Warning signs may be ignored or overlooked, therefore other precautions should be taken.

Play areas

Children can injure themselves by throwing metal wires or other objects over lines or flying kites near lines. Any overhead lines near play areas should be covered so as to avoid accidents.

Wildlife

Small birds sit on power lines, hunting birds sit on pole-mounted equipment and larger birds can fly into lines. Small animals (squirrels, domestic cats etc.) can run along and across lines. Large birds can also sit on cross-arms and touch the phase wires with their wings on take-off. With bare wires these birds or animals can kill themselves by shorting out the phases to each or to earth. In order to protect wildlife completely, the use of covers over bare end fittings and arcing horns will be required. Flying birds can often not see the overhead line in time to avoid it in their flight path, and so will be electrocuted and/or severely injured in the resultant crash.

On the road

There are other areas where people can be killed by touching overhead lines. The free wayleave at road edges means that many overhead lines follow the road verges. In road accidents, therefore, it is likely that a pole supporting an overhead line can be hit. The actual impact may kill or injure the people involved.

2.5.2.2 Fatalities when lines are damaged

Forests

Fallen trees caused by storms occur frequently in forests, and these can land on power lines. Non-professional people can sometimes start to remove these trees to avoid electricity failures, and this can often result in electric shock with bare lines.

Broken conductors

Downed conductors, either broken or close to the ground due to a lightning strike, pole failure or heavy snow loads, may not always touch the actual ground and operate the network trip. This type of incident can be a major hazard to people and animals and has been the cause of many fatalities.

On the road

There are other areas where people can be killed by touching overhead lines. In road accidents where a pole is hit, the line may come down very close to the road and it is possible that those people in the vehicle can be electrocuted.

2.5.3 Fatalities – accident statistics

Accidents at work and to the public are collated in the UK by the HSE. Over the ten years to 1997 these statistics have shown that there has been an average of 20 fatalities a year associated with the electricity supply system. There have also been around 400 injuries per year, approximately 250 among workers and 150 to the public. All these injuries and fatalities have been associated with bare wire overhead lines – mainly at 11 kV.

HSE records indicate the most common type of incident that results in a member of the public being killed. These are:

- car accident involving roadside ESI wood pole carrying an OHL
- microlight aircraft flying into the lines
- fishing poles touching an OHL (all lines crossing fishing areas have notices banning fishing, but sometimes these can be ignored)
- use of metal ladders or long metal pipes
- using vehicles with raised sections (e.g. JCBs, HIAB, cranes, tipper lorries etc.)
- hot air balloons touching the OHL
- people erecting 'zip-up' scaffolding near lines
- kite flying near lines
- raising aerials at campsites etc.

- fêtes where people carry long metal poles or tent supports near lines
- farm workers standing on hay on lorries.

These are all taken from actual incidents and they are unfortunately repeated year after year.

2.5.4 Higher voltage lines

Excluded from the statistics in section 2.5.3 are instances where aircraft have touched 132 kV overhead tower lines. At times these lines can be difficult to see when strung with bare conductors. They also represent a navigational hazard near the numerous small airfields around the UK.

2.5.5 Accidental third-party damage

Accidental third-party damage can occur to overhead lines in many situations similar to the above scenarios. The overhead line clearances stipulated in ENATS 43-08 allow for only 5.2 m over farmland and 5.8 m over roads (note that all Electricity Association standards are now re-titled Energy Networks Association Technical Specifications or ENATS). These values are backed by the Electricity Act and are a legal requirement. However, farm machinery has changed substantially since most of the UK overhead network was constructed and can easily exceed these clearances, as can JCBs on building sites. The accidental contact of tractors, bailers, JCBs etc. with overhead lines is, unfortunately, not uncommon. Maps of underground cables are often uncertain or even not referred to when the ground is excavated. Underground cables can be exposed and the problem then is to identify the circuit so that it can be switched off while re-instatement or repair is carried out. Another accidental third-party effect is when a road vehicle hits a wood pole support. This can break or cause the pole to lean, thereby allowing one span to sag close to ground level. The public may not appreciate that the line is live and potentially lethal.

2.5.6 Vandalism and theft

Overhead line conductors and underground cable can be a source of criminal theft. In order to steal the cable/conductor, the thief may deliberately induce a fault by shorting out the line using, for example, a metal chain. The line protection may come out or fuses blow, thereby rendering the system dead. However, the theft may leave the supply open and it may therefore be later re-energised assuming a non-damage fault. In this case the vandalised line may be live and injure or kill an innocent victim.

Porcelain or glass insulators protect the public from the high-voltage lines. These can be damaged or shattered by air rifles or shotguns in order to simply see the effect, allowing stay wires to rise to high voltage and people or animals to be electrocuted. The removal of anticlimbing devices (ACD) such as barbed wires can allow access to the high-voltage line to satisfy daredevil desires. Warning notices are often ignored with fatal consequences.

The utility has, therefore, to build and maintain a line that is resistant to accidental or deliberate third-party damage. The line must satisfy current safety legislation when built and also throughout its life. Section 2.7 details this legislation.

2.5.7 Mitigation of danger

The covered conductor (CC) overhead line option, in whatever guise, significantly mitigates the danger of introducing an electricity supply above ground. This system therefore works with the environment as it allows the overhead line to be introduced in areas where it can provide a safer environment for the public and wildlife. It will also assist in preventing interruption from supply, e.g. from objects blown or placed on the OHL conductors, to a greater degree than the bare conductor design. Safety aspects of CC use are covered in more detail in section 9.4.4.

2.6 Legislation and the environment

2.6.1 General

Some aspects covered in this section will already have been discussed. However, the environment is becoming more important to us all and it may be worthwhile to look at some areas anew from an environmentalist viewpoint.

It is important to understand that, in most cases, the likely major effect of an electricity overhead line is the visual impact on people who live, work, recreate and visit an area. Two options are available for reducing, but not eliminating this impact – the choice of components or design and careful routing.

It is important to recognise that under the terms of the UK Electricity Supply Act 1989, designers should consider environmental, technical and economic matters and reach a balance between them. This means that the proposed route will be the one, selected after an evaluation of a number of route options, that best fits the specified selection criteria.

2.6.2 General environmental issues

2.6.2.1 Route selection

The aim of route selection is to identify a technically feasible, reliable and economically viable overhead line route, between two or more specified points, which causes least disturbance to people and the environment.

Some of the environmental aspects that will need to be considered are:

- visual amenity, recreation and tourism
- landscape resource
- nature conservation
- agriculture
- the cultural heritage.

These points illustrate the need to provide flexibility in order to maintain a reliable supply to our customers at an economic level. Route selection has been covered in more detail in chapter 1.

2.6.2.2 Wildlife

It is a fact, however, that to re-design or introduce retrofit devices to the whole of a network to prevent, for example, large wingspan birds from landing on an overhead line cross-arm is not only uneconomic but, unfortunately for the birds, unrealistic. Insulator covers on transformers can reduce fatalities for hunting birds such as falcons as these often use these as platforms to look for prey.

2.6.2.3 Environmentalists

Although it is important for the electricity supply industry to work with the environment, there is also a clear responsibility to provide electricity to customers. This should be achieved by being safe, reliable, unobtrusive, accessible, maintainable and economic.

Getting the user and the environmentalist on board is a major hurdle for utilities. However there is no guarantee of progress until further hurdles have been cleared. The acceptance by the utility provider's staff and then the landowners and local authorities of where the overhead lines are to be built can often be the greatest stumbling block.

2.6.2.4 The sensible solution

It surely seems sensible for all agencies to work together:

- to address the needs of the utility provider in supplying a reliable supply to their customers
- to reduce the difficulties in gaining wayleaves for new and important overhead line networks in sensitive environmental areas at economically acceptable levels.

Any increase in initial construction cost to achieve the above aims must be justified. This justification can come if it can be shown that there is indeed both a safety and an environmental case for introducing or continuing to use overhead lines for certain routes. As a design engineer looking for flexibility and options, a variety of designs must be part of the available toolkit and considered as an option as and when the circumstances dictate.

2.7 A summary of UK legislation

2.7.1 General

This section details the legislation involved in the use of overhead lines in the UK. The full text of the material can be found on-line and in libraries. The legislation is presented here in brief schematic form to enable the reader to pick out a suitable subject area for further study.

2.7.2 The legislation

The electricity supply regulations in force before 1988 laid down in detail the requirements for transmission and distribution line design parameters including:

- line conductor materials
- minimum size of conductors
- minimum height above ground of line conductors
- minimum height of wires and cables other than line conductors
- stress limitation in line conductors and other wires and cables
- support and foundation design.

Major changes in safety legislation have been introduced since 1988 that affect overhead line distribution design, construction and maintenance.

The relevant pieces of legislation are shown below with the key pieces in **bold**:

a **The Electricity Supply Regulations 1988**
b **The Electricity at Work Regulations 1989**
c The Provision and Use of Equipment Regulations 1992
d The Personal Protective Equipment at Work (PPE) Regulations
e **The Management of Health and Safety at Work Regulations 1992**
f **The Construction (Design and Management) Regulations 1994**
g The Construction (Health, Safety and Welfare) Regulations 1996
h **The Electricity Supply Quality and Continuity Regulations (ESQCR) 2002**
i **BS EN 50341 and BS EN 50423**

These regulations provide a comprehensive reference. However, the main areas of interest are covered in regulations in ***a***, ***b***, ***e***, ***f***, ***h*** and ***i*** above.

Regulations 4(1) of the Electricity at Work Regulations 1989 requires that 'All systems shall at all times be of such construction as to prevent, so far as is reasonably practicable, danger'.

From 1988 onwards the principal requirement for safety is that the construction shall at all times prevent danger, so far as is reasonably practicable. It gives the supplier the responsibility of defining requirements in terms of detailed technical design.

The regulations also recognise that the construction will not retain its original condition throughout its working life and place responsibility upon the supplier to decide when the plant may give rise to danger. So there is a long-term responsibility. Therefore the designer has to bear in mind the deterioration caused to overhead lines by severe weather exposure e.g. near any coast and in upland areas (icing).

The Management of Health and Safety at Work Regulations 1992 continue this philosophy by requiring that, among other things, risk assessments be carried out on plant when the original condition is no longer valid. This is very important.

Another legal requirement that must be addressed is the Construction (Design and Management) Regulations 1994.

The principal objectives of these regulations are to ensure proper consideration of health and safety issues throughout every phase of construction from feasibility studies to demolition.

2.7.3 Electricity Supply Regulations 1988

The Electricity Supply Regulations 1988 do not go into such detail as previous – now redundant – legislation and the requirements for OHLs generally cover ground clearances to live conductors and precautions against unauthorised access to lines. The responsibility to construct, install and protect both electrically and mechanically is the responsibility of the supplier, as also is the use and maintenance of the network to prevent danger or interruption of supply. This is not the first time that this has been stated in this section but its importance cannot be overemphasised. Even if the supplier's design gets DTI approval, it is still the supplier's responsibility that it is safe and 'fit for purpose'.

2.7.4 Electricity at Work Regulations 1989

These regulations cover the following points:

- All systems, plant and equipment to be designed to ensure maximum level of safety. Systems include both permanent and temporary installations, i.e. construction sites.
- Installation and maintenance to reflect specific safety requirements.
- Access, light and working space to be adequate.
- Means of cutting off power and isolating equipment to be available.
- Precautions to be taken against charging.
- No live working unless absolutely essential.
- Specific precautions to be taken when live working is essential.
- All persons to be effectively trained and supervised.
- Responsibility for observing safety policy to be clearly defined for both employer and employee.
- All equipment and tools to be appropriate for safe working. Particular reference is made to testing portable electrical appliances in offices in HSE Publication IND(G) 160L.

2.7.5 Management of Health and Safety at Work Regulations 1992

The duties of employers are summarised below (note: provisions must be in writing if more than five people are employed):

- To carry out risk assessments when the original condition is no longer valid.
- Implement health and safety arrangements covering planning, organisation control, monitoring and review of protective and preventive measures.
- Provide health surveillance appropriate to health and safety risks identified by the risk assessment.

- Appointment of competent persons to assist the employers to discharge their responsibilities.
- Provide procedures for serious and imminent dangerous situations that may arise.
- Comprehensive and reliable information to be made available including identification of risk, protective and preventive measures, procedures for emergencies.
- Provision of staff training including when first recruited, upon transfer or change of responsibilities, the introduction of new equipment and technology or when new systems of work are introduced.

2.7.6 Construction (Design and Management) (CDM) Regulations

The object of these regulations is to:

- Ensure proper consideration of health and safety issues throughout every phase of construction from feasibility studies to demolition.
- Obtain better management and co-ordination of health and safety issues from preliminary design to handing over to client and eventually to demolition.
- Achieve the appointment of competent (from a health and safety standpoint) designers, planning supervisors, principal contractors and contractors.
- Create two instruments for managing and co-ordinating health and safety:
  1 the project health and safety file (which is a maintenance file with flagged health and safety issues)
  2 a health and safety plan relevant to the work being undertaken.
- Obtain an adequate allocation of resources and sufficient time to ensure that duties imposed by these and other health and safety regulations can be implemented.
- Place responsibility upon the client to make financial provision and adequate time available in the project programme for the implementation of health and safety legislation.
- Involve all participants in the achievement of safe project working environments, i.e. client, designers, planning supervisors, health and safety co-ordinators, principal contractors (the contractor for co-ordination of health and safety matters during construction) and all sub-contractors.

2.7.7 ESQCR (2002)

The Electricity Supply Quality and Continuity Regulations are the current UK ESI regulatory standard. They continue to treat covered conductors as bare wires, i.e. the insulating effect of the sheath is not taken into account, and so the medium voltage clearances are not changed from those given in Table 2.1.

The values in Table 2.3 are in accordance with ENATS 43-8, which will be followed in this specification.

2.7.8 BS EN 50341 and BS EN 50423

BS EN 50341 refers to overhead electrical lines exceeding AC 45 kV, and draft prEN 50423 refers to overhead electrical lines exceeding AC 1 kV up to and including

Table 2.3 Minimum height above ground of overhead lines (Electricity Supply, Quality and Continuity Regulations 2002)

Nominal voltage	Minimum height above ground, m[1]
Not exceeding 33 000 V:	
over roads accessible to vehicular traffic	5.8
all other situations	5.2
Exceeding 33 000 V but not exceeding 66 000 V	6.0

[1] The minimum height above ground of any overhead line shall be calculated at the maximum likely temperature of the line conductors.

AC 45 kV. The UK is obliged to adopt these new standards and in particular their view on wood pole design using probabilistic methods (called the 'general approach' in these documents). With respect to wood pole lines, however, the UK has decided to adopt the 'empirical approach'. These standards are covered in chapter 4.

2.8 Summary

This chapter has attempted to cover an essential part of overhead line work. Any change that is considered by the local or county authority to be significant (they all have their own rules) requires consent from several bodies. 'Significant' may include changing the voltage, raising the line height by more than one metre, changing conductor type etc. The best-designed line in the world is no use if permission cannot be granted to erect it. The stages covered have included the legislation concerning wayleave, safety and the environment, the very important Electricity at Work Act of 1989 and the Electricity Supply Regulations of 1988. The Management of Health and Safety at Work Regulations of 1992 and the CDM Regulations of 1994 complete the most important health, safety and responsibility aspects. Health and safety requirements are now an essential part of power line design and the OHL engineer must have a thorough understanding of current legislation.

2.9 Further reading

This chapter has covered the current electricity supply regulations. The new regulations can be read on http://www.dti.gov.uk/electricity-regulations:

'The Electricity Supply Regulations 1988'
'1989 Electricity Act'
'The Electricity at Work Regulations 1989'
'The Provision and Use of Equipment Regulations 1992'

'The Personal Protective Equipment at Work (PPE) Regulations'
'The Management of Health and Safety at Work Regulations 1992'
'The Construction (Design and Management) (CDM) Regulations 1994'
'The Construction (Health, Safety and Welfare) Regulations 1996'
'Fatal accidents in the farming, fishing and forestry industry, 1999–2000'. HSE publication
'Effect of current on human beings and livestock'. IEC 479-1, 1994
'Electricity supply quality and continuity regulations'. ESQCR, 2002
BS EN 50341: 'Regulations for overhead electric lines exceeding AC 45 kV', 2001
BS EN 50423: 'Regulations for overhead electric lines exceeding AC 1 kV up to and including AC 45 kV', 2003
ENATS 43-8: 'Minimum electrical clearances to overhead lines for voltages up to 400 kV'

Chapter 3

Surveying and profiling

3.1 Scope

This chapter looks at surveying and profiling – essential first requirements for any new line construction or for refurbishment. Traditional ground-based techniques are covered first, followed by a description of the increasing trend to use helicopters and laser profiling. The chapter is split into three main areas:

1 traditional surveying /profiling for a proposed new green field route
2 refurbishing an existing line
3 laser profiling using helicopters.

3.2 Introduction

Before any new line is built, or any major changes made to an existing line, it will need to be surveyed and profiled. Essentially, surveying means going along the line and making a detailed note of:

- every pole (height, age, condition)
- span length
- conductor type
- nearby tree growth
- land heights relative to the conductor
- land height relative to mean sea level (normally to a local Ordnance Survey (OS) datum point)
- road, rail, river or footpath crossing
- fences, houses etc. near to the line.

Indeed, anything that may be in any way remotely affected by the existence of the overhead line. Surveying is essential as new estates are built and land changes use in ways that could affect safety or normal operation of the line. The OHL engineer does not have to do this – there are specialist surveying companies to carry out the tasks. The output of all this will be a profile of the land within, say, ten metres either side of the existing or planned line route, which includes the clearance values under the conductor along each span.

It is important to realise what is involved and what information will be obtained. It is also important to know what information will not be obtained. Standard surveys and profiles are relatively inexpensive in comparison to line construction or refurbishment costs. But, as with anything standard, it is wise to know that extra non-standard requirements may cause the price to increase substantially. It is no good giving the surveyor a route and then asking whether it would be better 50 m to one side or round the other side of that hill. As the cost of a line has to include surveying and profiling, the cost needs to be kept down in every aspect.

The surveyor also needs some things from you. For a proposed new route the surveyor needs:

- the proposed route with agreed planning permission
- a right to survey (wayleave) from the grantors (normally the landowners).

If an overhead line actually exists, or if a parallel route within 25 m needs to be investigated, it is only necessary to contact the landowner/occupier before starting. Although the landowner will normally have no legal rights to stop the survey, it is as well to provide advance notice, especially if there are crops along the route. It may also help the surveyor to have prior knowledge of any sensitive or dangerous animals that are likely to be present on the route.

3.3 Profile and survey – the traditional way

3.3.1 Profile and survey needs

The OHL engineer can appreciate the need for an electricity supply and will also have a good idea of how he wants to design it electrically. The basic design standard to which it will be built will also have been decided. The approximate route will have been worked out, but what then? There are more stages yet before the engineer can give the plans to the linesmen to build the line. Pole positions can be detailed mechanically for the best line design, but will they annoy the farmer? If so, maybe an accommodation can be arranged where poles are placed conveniently in a hedge or other field boundaries. Normally most farmers will be co-operative, and the survey will identify hedges to hide the poles in. But can the design accommodate a long span across a large field to match up with these pole positions? Are angle/section poles required? Is there a road that can be crossed obliquely or is another pole necessary to cross at right angles (as for wide dual carriageways)?

Ground conditions along the route also need to be known, whether it is a new line or a refurbishment of an old one, as the last thing required is a surprise. Maybe someone has built a hen house on the proposed route or there is a well-used footpath that is not marked on the maps or there are other places to avoid etc. Ground conditions can change – from arable land to a local swampy pond to an outcrop of rock or whatever – so foundation design may change. Information on the access situation for vehicles along the route could also prove useful.

3.3.2 Maps required

Several maps are required and they all help to solve problems quickly and effectively. At times, fine detail is needed, and on other occasions the global view is what is required. An Ordnance Survey (OS) Landranger (LR) map has a scale of 1:50 000 and is useful to get a general idea of the route and to see where the footpaths and contours go from a general perspective.

The 1:10 000 scale map is, however, the workhorse for existing routes. The overhead line engineer marks the proposed route on this in detail; it is also handy for planning longer routes and access points.

The 1:2500 scale map is essential for locating new pole positions, identifying OS features and scaling from known points for a new route. These days, however, access to the internet (e.g. www.streetmap.co.uk) allows the direct digital generation of the type of map required rather than relying on shop-bought maps.

Then, of course, the survey needs to be precisely located with respect to the rest of the UK. This is undertaken by relating all surveys to a local OS datum benchmark (BM). For long routes (>5 km) the survey may be related to more than one BM. These BMs are marked on 1:10 000 maps and can be seen as marked plaques on an established building in a town or city. They are not to be confused with the OS trig points on numerous hilltops throughout the UK.

Today, the use of helicopters inspecting existing lines has enabled pole positions to be precisely located using the global positioning system (GPS). This can be linked with digital maps to allow more precise planning than traditional methods allowed.

3.3.3 The surveying and profiling work

The surveying and profiling work will now be described briefly. Traditionally, the surveying team has needed to carry the following load of equipment:

- theodolite
- field computer
- profiling software (or equivalent)
- tripods
- prisms
- prism pole
- 12 one metre ranging rods
- pegs
- fluorescent tape
- spray paint
- two-way radios
- mobile phone.

The final data will normally be delivered on a disc or CD, but some OHL engineers still prefer the traditional paper copy, especially for off-site work. The development of laptop computers for field use has provided what can now be a preferred option, and these also allow a greater availability of information on surveys and equipment.

3.3.4 Traditional survey methods for a proposed green field route

The object of overhead line surveying is to relate all points and features along a linear length to one another. In effect, this means that although the line route may bend (angle) at various points, it is useful to have a straight linear survey of the strip of land that is of interest.

So initially the route must be established. This means that the planned line proposal must be marked out on the ground. Normally, the terminal or tee-off point at the source end is located by scaling from a 1:25 000 plan feature such as fence line/wall etc. Alternatively, if digital maps have been made available, co-ordinates can be noted for all the angle and terminal poles in the office. Then, provided co-ordinates are given for some existing features, the electronic distance measuring equipment can be used to orientate the line. This pinpoints the proposed angles and terminals and fixes the line route – but not the intermediate pole positions. This method is more suited to GPS surveys that are now becoming more widely used within the industry.

It is then necessary to peg out all the straight parts of the route, i.e. the lengths between angle poles. A prism is positioned above each angle point and the instrument is set up at a vantage point anywhere, to sight and measure the distance to both. The instrument then uses trigonometry to calculate the distance between the two angles. The position of the instrument, and any subsequent readings between the two points, enable the surveyor to guide the assistant to any point with a zero off-set. This process is repeated until the entire the route is established.

These traditional methods were used until the mid-1960s. A theodolite would be used to establish the route, a chain to measure the horizontal distances etc. A chain is a specific length (commonly 100 ft) that is not susceptible to stretching and so maintains the correct distance. The theodolite was used to fix the line peg positions – a task that was made easier by experience. Hedges impairing the view were scaled from the map.

Surveying with tacheometers became more widespread. Tacheometry is the indirect measurement of distances by optical methods instead of using tape or chain. The most important advantage of this method is that the fieldwork can be carried out much faster than with the theodolite, level, levelling staff and chain. The chaining of the route is eliminated and the levelling speed greatly increased. This is due to the fact that the tacheometer has a range of sight of the order of 700 ft and readings can be taken with the telescope at any vertical angle necessary.

In the previous method, the level had to be truly horizontal, so that its range on sloping ground was very limited, but with the tacheometer, a range of 700 ft in both directions along the route line is available. This means that in favourable conditions a maximum of four settings of the instrument per mile is required. This is important as damage in arable fields is reduced to a minimum, so avoiding annoying landowners who may refuse future planning permission.

3.3.5 Downsides to tacheometry and the modern approach

The downside of tacheometry was that it generated more involved office work. Modern methods of recording levels and chainages have become much easier. The first task is to transfer a level from the nearest benchmark to the start point, and then, using the data logger and computer, the proposed route can be levelled. Levels are taken along the route centre line along with the chainages. These instruments used in conjunction with satellite navigation systems allow measurement of the slope distance, vertical and horizontal angles etc. Levels are taken along the centre line at all points where the difference in altitude is in excess of 300 mm and at all obstructions, e.g. roads, railways, rivers, walls, ditches etc. Levels are also taken at all obstructions up to 30 m either side of the centre line. The width of this corridor depends on the voltage/size of the line to be built.

All obstructions are recorded with a ground level and height, and the dimensions are automatically computed so that the survey is fully compatible with both 2D and 3D images. Where the ground slopes across the line route, the level of the ground left and right of the centre line (CL) is also recorded at the following distances:

1 kV pole design	3 m
33/66 kV pole/tower design	5 m
132 kV tower design	7 m
275 kV tower design	9 m
400 kV tower design	13 m

Off-set levels not exceeding 80 mm difference from the CL are not recorded.

All trees are recorded within the corridor with a ground level and height. The surveyor also notes the species, height, trunk diameter and whether it requires lopping, felling or is okay.

Full details of power lines and telecommunications lines are also recorded.

If the proposed pole/tower positions are known, i.e. angle terminal, pole transformer equipment or tee-off, special care is taken to ensure sufficient levels are taken at these points. Angles of deviation are measured at each angle position and tie-in sketches made relating the pegs to at least three features shown on the OS plan. In remote featureless areas, reference points, bearings and distances are of more use.

3.4 Surveying to aid refurbishment of an existing line

3.4.1 Existing pole line surveys

Existing power lines are surveyed in much the same way without the need to establish the route with line pegs. All poles are recorded by giving a level alongside, measuring the height of the top and the conductor attachments (generally the lowest attachment point is sufficient). The pole number details and any equipment are noted along with visual checks on the condition and a hammer test. The hammer test is when the linesman hits the pole in several places up to 1.5 m from ground level. The sound of a solid pole is distinctly different from that containing substantial rot. The test however, is dependent on the experience of the linesman and it is difficult to estimate the extent of any rot present. The test may therefore result in poles being condemned when they still have significant economic remnant life left. Stays are also recorded by length, size and position, and shown on the tie-in sketch.

The condition of each pole may be required more accurately than by simply using a hammer test. The surveyor can therefore be asked to use a mechanical or an ultra-sonic probe to detect rotten poles and estimate the remaining strength of the others. Specialist firms can supply these data now more efficiently than the traditional surveyor. This is covered in more detail in chapters 12 and 15.

3.4.2 Existing tower lines

The same principles are applied when surveying along the route of an existing tower line. Each tower is recorded by giving a level in the tower centre where the surveyor will then follow the instructions of the tower monitoring if using the commercial software *or* sight the bottom centre brace followed by the peak and record both heights. A level will then be taken and the reference number recorded.

Alternatively, a level on the lowest concrete top can be taken, then a reading beneath each bottom cross-arm, followed by a height of the underside of the cross-arm and the conductor attachment point at both ends of the bottom cross-arm. The recording of levels and obstructions is the same as for a pole line survey. The off-sets are measured from the aerial earthwire. Great care must be taken to ensure that sufficient information is collected, bearing in mind the bottom conductor could be as much as 10 m to the left or right of the earthwire due to the tower construction or local winds.

3.4.3 Profiling data

After each day the survey data for existing or proposed lines is downloaded from the field computer either onto a floppy disc or direct into a PC. For the PC to process the data, it must be loaded with either profiling pole CAD or tower CAD software. Such commercial software may be available, but in most cases specialist firms use it on a sub-contract basis.

The data are then edited and the level checked to see if they fit with the benchmarks. The features are checked for correct terminology, the remarks for clarity and the land use for being correctly documented. At this stage, the data are displayed in a spreadsheet format with feature description, chainage, off-set level and remarks.

Pole line data are normally in lengths of up to 2.5 km in each file, with the chainage split at each angle position. Tower line data are split at every tension tower, thus creating files on odd occasions in excess of 5 km.

The data are now ready to be modelled. This is a process done within a micro-station which creates a 3D drawing file known as a .DGN. Within this file, the section of survey can be viewed as a 2D file in elevation scaled up to 1:2000 horizontal and

1:200 vertically, or in 3D unscaled (1:1) in elevation and plan view. If there are any anomalies these will show up as the .DGN file is viewed and checked; the file can be modified by returning to the editor program and amending the data, followed by remodelling the section of the survey.

Once the .DGN files have been checked and all is in order, design work can commence. Prior to this, the tie-ins are created within a CAD environment. This is a process that involves copying the surveyor's field sketch, clearly marking the dimensions and features. These are measured along with the angle of deviation and then the sketch orientated as per an OS plan. This makes everything easier to quantify. The spare set of route plans is then marked up using the surveyor's tie-ins and the chainages. The overhead line route is finally plotted.

3.4.4 Digital mapping

With the use of digital maps becoming more widespread, a growing percentage of plan marking is done on the PC. A pole line survey is recorded in a different format to tower line routes, enabling a 3D copy of the centre line survey viewed in plan view to be superimposed directly over an OS digital plan. Fixing on the start point, the subsequent angle positions should fall correctly into position, provided that all angles of deviation have been measured correctly (inevitably a little massaging is necessary) – a ten-minute error over 2 km equals 5.8 m.

It may be that tree cutting is required. This can be planned and scheduled from the survey data. The relevant detail is copied from the data and correlated to form a schedule comprising section, species, chainage, height, diameter and remarks, e.g. lop, fell or okay.

3.5 Traditional profile plotting and line design

3.5.1 Profiling

With the survey done the profiles can now be plotted. In commencing profile plotting the level notes are examined to ascertain the range of levels so that the profile can be suitably placed on the paper. If the lowest level is 410 ft, little purpose would be served in measuring this distance vertically, so a starting point of 400 ft Ordnance datum (OD) would be assumed.

The reduced level of the starting point is plotted and then the other points at which the levels have been taken. The profile is drawn through these points, and any hedges, walls etc. are shown. The proposed structure and the bottom conductor attachment points are then added to the profile using made up templates. Having completed the profile, the next stage is to apply the sag templates (see chapter 7 on sags and tensions). The object is to design a power line that meets the specification criteria and the landowner's requests.

The regulatory clearances must be met throughout the lifetime of the conductor at the maximum continuous operating temperature. This is normally 50 °C but some companies wanting more out of their assets require the lines to be able to operate at up to 80 °C. The profiles will show up to four curves. The first is the conductor curve at the tension required after allowance for creep and at the temperature requested – the hot curve or hot template (see below). A second curve is plotted parallel to the first but reduced in height by the statutory (or requested) clearance value. The third curve is the support foot that represents the actual height from the base of the standard support to the point of attachment of the lowest conductor. With single supports, such as wood or concrete pole lines, the support foot curve is not much used. This is because the supports are, where practicable, placed in hedges or other suitable places. However, with tower lines, this curve is useful for indicating the actual tower positions. The fourth curve is the uplift or cold curve (cold template) at −5.6 °C in still air.

The hot curve is of itself little use, but the parallel curve (or the use of a sag template – see section 3.5.2) with the requested clearance taken off is of vital importance. This shows immediately whether the design parameters will give sufficient ground clearance. If they do not (or if they give too much due to the ground falling away between the pole positions) then the pole heights or conductor tensions may need to be adjusted. This process goes on until the curve clears the ground in all places.

The cold curve is of use as it indicates whether the force applied by the conductors on the poles will be a downpull (as is normal) or an uplift as the cold temperatures cause the conductor to reduce in length. This can be a problem with uneven land where the two poles on either side are higher than a central pole. If the cold curve shows that the conductors change from pulling down on the pole to pulling it up out of the ground, then this uplift force needs to be considered in the line design (chapter 5). Figure 3.1 shows a typical set of curves.

These curves are the initial high-tension curve on a new conductor when erected at a typical ambient temperature of around 15 °C. After creep (geometrical settling down of the conductor strands and metallurgical alignment of the conductor material)

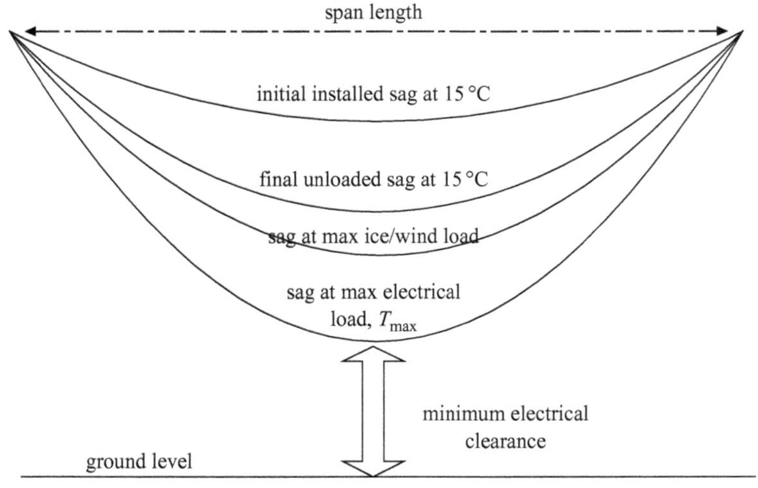

Figure 3.1 Sag curves

the unloaded sag curve is produced. Creep is discussed in detail in chapter 7. Allowance has to be made for potential snow/ice loads and the maximum temperature expected under normal (i.e. not short-term fault conditions) maximum electrical conditions.

3.5.2 *Use of the sag template*

The method of applying the sag template is as follows.

The template is placed horizontally on the profile and moved until the 50 °C line coincides with the tops of the single supports already drawn in at normal heights. If any intermediate points between the supports on the profile are cut by the groundline on the template, the support or supports require extending in order to give the necessary ground clearance. Alternatively, the conductor tension could be increased – but this will affect pole loading and other parameters. With tower lines, the template is applied but the support foot line indicates the best position for the tower with the maximum span permissible on the type of construction in use.

It may be necessary to use extended towers in some instances due to the fact that angle tower positions are selected on the original survey. The cold template or −5.6 °C uplift curve is then applied. The template is again applied horizontally until the tops of alternate supports coincide with the −5.6 °C curve. If the curve is above the intermediate support, this support is extended until it touches the uplift curve and so eliminates uplift. It will be appreciated that the applying of the template is a job that must be done carefully and, in the case of tower lines, the total cost of the line will depend on the judicious application of the template. Any wrong assessment of a support height or any infringement of ground clearance may result in uplift, which may not be detected until conductor erection.

The line profile is now complete and the line only requires to be built. The profile is now used pole-by-pole (or tower-by-tower) to establish all the pole-top hardware needed and the lengths in which the conductor will need to be ordered. This is known as the schedule and is a vital requirement for planning and construction.

The purpose of pole and tower schedules is twofold:

1 it is the basis for the ordering of poles, towers, insulators and conductors etc.
2 it is the engineers' guide on construction.

3.5.3 *Design using a software package*

Automated design work is one of many applications available within a software package. Once the system has been set up, it needs to be configured to suit the line specifications desired. Each utility will have its own individual specification.

The procedure is straightforward:

1 The conductor parameters are entered to enable the design sag to be created.
2 This is followed by the creation of a pole database. This can be very simple, containing the range of poles for terminal angles, intermediates and tee-offs, or configured fully so that all the relevant fixtures and fittings are noted enabling a complete material schedule to be produced.

3 The first process after modelling is to choose the minimum size structure that is available for each different type of support, e.g. a 10 m size pole may be the minimum allowed for a terminal.
4 These poles are subsequently added to predefined positions that have been located in the field, e.g. transformer locations and tee-offs.
5 With these in place, the autopoler is run which, in brief, will place poles on the design at preset design criteria, this being previously loaded into the PC in various database files. (With respect to the autopoler, this will only place poles at the optimum span length between defined positions so, although it may be adequate to have three spans between a term and angle, these positions may require amending if it places a pole in the middle of a prime arable field.)
6 Once the autopoler has completed its task, the designer should then carefully check through its suggestions, moving poles to boundaries where possible and reasonable to do so.
7 The autopoler is then re-run to plot the remaining poles.
8 Once the poles are placed, the strengthener is run across the design to give a suitable diameter to the previously height-only poles. Pole strength is related to its diameter and a pole is therefore specified by its height and diameter one metre from the base. A schedule can now be produced showing pole types, sizes, span lengths, land ownership and specification number.

A further step in the process can be performed by running a separate scheduler across the design, which will provide a comprehensive materials list down to the last nut and bolt. This materials list is essential for planning the construction schedule.

The design is then checked to ensure that the preset criteria have all been met, i.e. wind loading, strut loading, section lengths and downpull forces.

Extra information taken from field notes can be added to the database files. This could include stay limitations (e.g. where there is only enough room to have the stay at, say, 30° instead of the optimum 45°). Assuming all these conditions have been met, then the section is ready to be put into a standard sheet format with the previously drawn tie-in sketches locating the angles, tee-offs and terminal. These are shown together with crossing sketches of telecommunication and different voltage lines. This profile creation is done to provide a hard copy in addition to, or instead of, the disc so that clients without the benefit of a PC are able to comment on the profiles.

3.6 Laser profiling

3.6.1 General

Helicopter inspections are becoming very common these days. Although expensive in terms of an hourly rate, the route length covered by a helicopter survey per day is far in excess of that by a surveyor on foot. Technical advances in lasers, digital video and GPS have enhanced the data that can be obtained from such a survey. Perhaps the main area in which the person on the ground is better placed than the helicopter is in wood pole condition assessment for existing lines. Topographical features, conductor height, span lengths, pole positions and land profile are all far more easily and quickly determined by a helicopter (or small plane) survey.

3.6.2 Laser scanning techniques

The most common laser scanning technique used today is known as LIDAR and this, combined with differential GPS, can capture accurate topographical data in a line survey. Specifically, the LIDAR system determines the line attachment points (for existing lines) and sag information as well as the condition of the right of way (ROW) for vegetation clearance. This last point enables tree-cutting schedules to be determined on an accurate conditions assessment.

3.6.3 Data gathering and output

The proposed route, or existing line, is provided to the helicopter team. This is specified on 1:10 000 Ordnance Survey maps. Other elements to come out of the survey will be ground and surface geological conditions and items that may have environmental impact.

The data are obtained by:

- aerial laser (this records the heights of all ground-based features)
- aeronautical INS (inertial navigation system) and airborne GPS (to define position of features)
- identification of local GPS base stations
- digital video camera.

The data are processed using commercial software packages to produce an overhead line (for existing lines) or route profile. The output can also include engineering clearance drawings, vegetation survey, thermal view and digital mapping of the route. The thermal view, using infra-red thermography, which can be overlaid onto real-time video pictures, can be used to pick out faults, such as poor connections due to the temperature rise.

Environmental factors such as recent buildings, streams or rivers, recreational areas, farm storage, animal occupancy etc., can also be identified.

3.6.4 GPS

These days virtually every overhead line support structure has been given a GPS location. The accuracy of GPS is dependent on initial local point determination, satellite availability and locality.

3.7 Further reading

MORECOMBE, W.: 'Overhead power lines' (Chapman and Hall)

SMITH, S.: 'Study of overhead distribution lines and their design parameters' (Energy Networks Association)

Chapter 4

Traditional and probabilistic design standards

4.1 Scope

Chapter 2 looked at the legal framework including acts, regulations and technical standards within which the electricity companies and hence the overhead line design engineer has to work. Chapter 3 covered the surveying and profiling of a line or route to help decide where it should go. This chapter looks at the technical standards to which the line needs to be designed. Here the problems of wayleaves and consents have been assumed to be finalised, and it is the technical details of the various components in the overhead line that need to be investigated.

The Electricity Council [before its demise in 1991 to become the Electricity Association (EA) and eventually (2004) the Energy Networks Association (ENA)] had to interpret the requirements of the acts and regulations in specific technical details and to standardise practice as far as possible throughout the United Kingdom. The usual procedure was for the Electricity Council to form a working party comprising area board members with specialist knowledge of a given subject (now known as distribution network operators or DNOs) to produce what was known as an electricity supply industry standard and now re-titled Energy Networks Association technical specifications or ENATS. Much of the work that was done to build up the comprehensive list of overhead line standards that we now have was begun before 1991.

This chapter, therefore, covers the basic design standards for overhead lines in the UK and will also touch briefly on the effect on the UK of European standards bodies such as CENELEC. It will go through the historical basis of traditional (or deterministic) line design as was universally used up until 1988. At this point, a new semi-probabilistic method was adopted. This trend is mirrored world-wide as more countries start to move over from traditional to a more probabilistic design basis.

The history of line design is still visible in the UK, as many lines still exist from pre-World War II days before the industry was nationalised. However, the vast majority of 11 kV lines in use today are based on designs laid down just after the war, from 1947 onwards. Bearing in mind that an overhead line has a lifespan of around

60 years, it is perhaps surprising that many of the old lines are still giving good service after so many decades. So the main point of this chapter is to:

- summarise how overhead line design has developed over the past 120 years in the UK
- explain the difference between probabilistic, load-factored and deterministic design
- recognise the advantages of looking at design and material properties from a probabilistic standpoint
- discuss the main technical standards applied to current overhead line design
- look at the software packages available to design components for overhead lines in specific weather environments.

4.2 Traditional design standards

4.2.1 Introduction

To understand why certain designs and standards were used in any given period of time it is necessary to look at the legislation that was current at that time. In this way, the reasons for certain designs become clearer. The overhead line standards used in electricity supply divide naturally into four groups:

1. period prior to nationalisation (1947)
2. from nationalisation to 1970
3. from 1970 to 1988
4. from 1988 to present.

4.2.2 Period prior to nationalisation (1882 to 1947)

This will not be a detailed history of the legislation, but enough will be given to know its general pattern and the major milestones as they affected the production of standards. In 1882 parliament was persuaded that legislation was necessary to promote and regulate the new technology of electric lighting. The first of the electricity supply acts was passed in that year. In the absence of any previous experience with electricity, much of the legislation was derived from the gas supply industry.

In many cases the driving forces for legislation are political and economic issues. The acts concentrated on organisational, legal, fiscal and administrative aspects but contained little in the way of technical and engineering information. Do not forget that in those days the industry was a growing collection of private firms with lines designed by manufacturers rather than supply companies.

The Electricity Regulations 1937 were the first regulations composed in a form that we would recognise today. However, in relation to overhead lines they gave little information of the type now known as standards.

So far as standards were concerned, individual companies developed their own designs to suit their own needs and, perhaps predictably, the numbers of line types were many and varied. Many of these designs are still there for us to see as several of the lines built in the 1920s and 1930s are still in operation today.

4.2.3 From nationalisation to 1970

Whatever the political or economic arguments for or against nationalisation, there was certainly a very strong technical case for it. The electricity supply industry in England and Wales comprised over 500 separate undertakings owned by private companies, local authorities and municipalities. From this number The Electricity Act 1947 created 12 English and Welsh Area Electricity Boards and the Central Electricity Authority. The date fixed for the transfer from private companies to the UK government (vesting of assets) was set by The Electricity (Vesting Day) Order 1948 for 1 April 1948.

Before this point technical standardisation among the private companies was almost non-existent, and although standardisation was highly desirable in order for the system to meet the increasing growth in demand for electricity, there was no single body within the electricity supply industry charged with the responsibility for standards.

However, from the British Standards Institution (BSI) came one of the first truly national standards to be used in overhead lines – BS 1320:1946 'high voltage overhead lines on wood poles for line voltages up to and including 11 kV with conductors not exceeding 0.05 sq.in.'. This design was both successful and timely. In the early 1950s the government promoted a countrywide programme of rural electrification and the BS 1320 design was admirably suited to the purpose. It was widely adopted and became the backbone of the UK's 11 kV light line system for the next thirty years.

4.2.4 The Electricity Act 1957

The next piece of legislation that had a significant bearing on overhead line standards was the Electricity Act 1957. Under this Act the Central Electricity Authority was divided into two parts:

1 the Electricity Council
2 the Central Electricity Generating Board (CEGB).

The CEGB now took responsibility for higher voltage lines (then 132 kV and above but later only the 275 and 400 kV lines) while the Electricity Council took overall responsibility for production of standards and introduced a system of working parties which influenced the design and specifications of OHLs.

4.2.5 From 1970 to 1988

The greatest influence on overhead line designs and standards during this period was The Electricity (Overhead Lines) Regulations 1970. These regulations contained a very considerable amount of technical detail. Consequently the Electricity Council system of working parties was most active during this period and produced the bulk of the overhead line standards that are listed in section 4.4. Key sections of these are given in Table 4.1.

Table 4.1 Extracts from The Electricity (Overhead Lines) Regulations 1970

Schedule 1	
Part II, Section 4	written authority granted by the secretary of state is required before any overhead line may be installed
Part II, Section 5	every line conductor shall be made of copper, aluminium or steel or any alloys thereof, or any combination of any such materials
Part II, Section 6	every line conductor shall have a cross-sectional area of not less than 12 mm^2
Part II, Section 7	height of line conductors above ground level is specified for various voltages
Part II, Section 9	stress limitations giving factors of safety of 2.0 and 2.5
Part II, Section 10	supports shall be of wood, steel, reinforced concrete or pre-stressed reinforced concrete or any combination of any such materials. Wood and steel shall be protected against decay or corrosion. The factors of safety on supports to be 2.5
Part III, Section 12	every electric line, support, wire and cable shall be properly and efficiently maintained
Part IV, Section 16	there shall be kept affixed to any support carrying a high-voltage line conductor a notice inscribed with the word 'Danger' in white letters of at least 30 mm in height on a red background, or in red letters of the same dimensions on a white background
Part IV, Section 17	every support carrying a high-voltage line conductor shall, if the circumstances reasonably require, be fitted with suitable devices so as to prevent, so far as may reasonably be foreseen, any person from having access to any position which is dangerously near any such line conductor
Schedule 2	
Part I	wind pressure on conductors
Part II	wind pressure on '*augmented mass (of ice...)*'

The nature of the information contained in the 1970 regulations is worthy of closer study because, although these particular regulations were replaced in 1988, much of the technical information is still relevant.

From Table 4.1 it can be seen that the amount of technical detail gives the standards engineer and subsequently the line design engineer virtually all the basic information needed to design an overhead line.

By the end of the 1980s, in addition to the variety of prenationalisation line designs, the standard wood pole overhead line designs were as follows (refer to section 4.4 for full titles):

- low voltage
  BEBS L1
  ENATS 43-30
  ENATS 43-12/13/14

- medium voltage
  BS 1320 11 kV light duty lines
  ENATS 43-10 11 kV light duty lines
  ENATS 43-20 11 kV and 33 kV heavy duty lines
  ENATS 43-50 132 kV Trident line.

As was often the case throughout the ESI, several of the boards would adopt one of the above standards but would then make modifications to suit their own particular operational, geographic or climatic circumstances. Thus there were many variants of the above list throughout the United Kingdom.

4.2.5.1 Overhead line design and the 1988 regulations: deterministic, probabilistic and load-factored design

General

Up to this point, line design had followed deterministic methodology. In the early 1980s, after the major snow/ice incidents of 1981/2, an investigation into alternative methods was started. This resulted in a technical document ENATR 111 that laid down the basic design principles following a semi-probabilistic methodology. In 1988 this concept of probabilistic methodology (in fact a semi-probabilistic methodology in this case) was introduced in the specification ENATS 43-40 which encompassed the work of ENATR 111.

Traditional deterministic overhead line design has involved the calculation of mechanical loads and deflections based on the assumption that lines are statically determinate. Deflections of conductors in wind have been taken as the same as those on a pendulum of equivalent mass and area to the span of conductor, with a length equal to the sag of the conductor. Loads transferred to the support have been thought of as being the vector sum of the weight and wind span of the conductor with the tension along the line of action of the conductor, all under particular loading conditions. These assumptions underpin the mathematical models behind the traditional deterministic design approach.

Probabilistic design is based on the recognition that nothing can be measured with absolute certainty, in particular, neither the strength of a component nor the applied load. In general, the actual strength and applied load will vary about a particular nominal value in accordance with a probability function.

The load-factored approach uses a more or less rigorous probabilistic analysis of applied loads and of anticipated conductor deflections together with a more or less traditional approach to the ultimate strengths of components and a geometric, partial-factored approach to set reliability levels.

Assumptions

There is nothing wrong with or intrinsically superior about either the deterministic or probabilistic approach. Each has limitations and each is based upon certain assumptions, but the deterministic base appears to align with an externally regulated regime, and the move towards probabilistic seems to align better with self-regulation. In practice, neither the traditional deterministic nor the more modern probabilistic design as applied to wood pole overhead lines reflects a purist approach. Both have been properly tempered by the practical consideration that is the hallmark of engineering design.

Deterministic design

Under set conditions of mechanical loading (the design loading case) the stress in each component has been limited to a certain proportion of the nominal ultimate capability of the component defined in terms of elastic limit or ultimate tensile strength. The factor by which the permitted stress is below the ultimate capability defined in this way is known as the factor of safety.

Under set conditions of conductor temperature, the clearance is specified between positions where a member of the public may stand (or park a vehicle) and uninsulated, live, conductors. Initially, conductor temperatures were also set or, more precisely, the temperature at which clearances should be maintained were set out, but in the 1970 regulations the term ‘conductor likely operating temperature’ was introduced to reflect the tendency to operate conductors at ever higher temperatures. No criteria were set for clearances at any other temperature, for example at 0 °C with ice loading. Under this situation conductors could meet the regulations at 0 °C with a high ice loading when barely above ground level as long as at a higher specified temperature the specified clearance was provided.

Over time it was found that lines built to these criteria performed well in some parts of the country and poorly in others. As a consequence of this, and as a result of the relative influence of each electricity board as regulations were reviewed, changes were made which were more or less arbitrary. Some electricity boards adopted designs that were much more conservative than the requirements of the regulations and others took advantage of the relaxations that were being made through the regulations. The effect of these trends was to cause a drift between the objectives of the deterministic models and their use. Over the years the design criteria had been relaxed to reduce the cost of rural electrification. Unfortunately, major failures due to snow and ice storms occurred every two or three years. The occurrence of two storms in the same year (1981 April and December) caused questions to be asked in parliament and an enquiry was set up. By 1985 the Baldock Enquiry had concluded that the process of relaxation had gone a step too far to be consistent with the requirement to ensure a sufficient supply of electricity. An analysis of nationally reported severe storm disruptions was conducted to show that lines designed to a probabilistic approach would have performed better than lines to the earlier BS 1320 and later derivatives, and, on this basis, withstand and reliability factors were specified for general application. The enquiry thus recommended that alternative design approaches should be investigated in an attempt to find one that would improve the reliability of overhead line systems.

Probabilistic design

This new approach allowed for additional local knowledge factors to be considered either externally by use of high wind factors or internally by varying reliability factors.

This reflected the relative importance of the circuit when a local case could be made against the 'fitness for purpose' clause of the regulations.

True probabilistic design was just coming into vogue for steel tower work but was ruled out for wood pole OHLs because of the different probability functions that applied to steel and wood in tension and compression. A new form of design termed load-factored design was introduced for the purpose.

In many cases the probability function will take the form of a normal distribution and most familiar probabilistic design is associated with normal distributions. However, the principles of probabilistic design apply equally well to all other probability functions. Underpinning such design there is the expectation that a reasonable estimate can be made of the probability function that applies.

In practice, failure is likely when the applied load exceeds the actual capability of a particular component to withstand the load applied. There is a distinction between failure in practice and failure to meet the design. The layman may anticipate that the design will ensure no failure in practice, but this is not realistic. There may well be circumstances that arise in which the device fails in practice while meeting the design requirements. This can occur, for example, if practical loads exceed those catered for by the design or if the actual strength of a particular component falls outside the range catered for in the design.

Probabilistic design accepts that at the limit there will be an overlap between the probability distribution of applied load and the line capability, yet if this overlap is kept small enough then the risk of failure is small. Of course, the effect of diminishing the overlap is to increase the margin between the centre of the two distributions, and this is generally associated with increased cost and less practical design.

One of the major problems with probabilistic design in the context of overhead lines supported on wood poles is the wide variation of performance of components – especially of poles and foundations. The situation is greatly confounded by the wide variability with time of applied loads. This situation is not nearly so acute in the case of overhead lines supported on steel towers where both the materials and the structures are more predictable. Nonetheless, it is important to recognise that with probabilistic design it is much more difficult in the event of a particular failure to determine if this is a failure to meet the design or one of the failures that the design accepted would occur.

A particular probabilistic design may accept a risk of failure of, say, 5 per cent of supports every 50 years. This may be based on the mechanical loads that may occur during likely weather conditions and the probability distribution of the strength of wood poles erected *in situ*.

It is clearly highly unlikely that following a particular event it would be possible to determine if a support which had failed had met or had not met the design requirements for the following reasons:

1 Such a support could have been a support that had strength at the lowest end of the representative probability function. We could not know this, however, since it would not have been possible to measure this *in situ* prior to the failure. Even if we were able to assess this after the event this would give us no more certainty,

not least because the pole will be broken at its weakest point and the foundations will have been modified by the failure.

2 Such a support may have had lower strength than was anticipated by the probability function. We cannot, however, know this with certainty since we would need to know the strength of all of the other poles as, at the limit, all probability functions allow for units of very low strength with decreasing likelihood. If it is found after the event that the pole exhibits rot it may be taken that it was weaker than it could have been. This is true but irrelevant in the assessment of the adequacy of the design since it may still have had strength remaining as described by the probability function.
3 The weather experienced may have been more extreme than had been catered for by the design.
4 The manner in which the mechanical loads accrued in light of the weather may have been more extreme than was catered for by the design.

Consequently, to assess the relevance of particularly a probabilistic design to practical considerations it is necessary to take a wide view of a series of failures over a period of time. These can be obtained from the National Fault Information Recording Service (NAFIRS). All utilities are required to report faults to this service.

Principally, the design uses more traditional calculations based on sets of values that use a probabilistic base to reflect the likely variation of loadings to be expected across the UK. In other words, the UK is now accepted as having several different areas that suffer different wind and ice loads, i.e. they therefore have different probabilities of suffering high stresses due to the weather. Some line components are the same across the country. A piece of steel is as strong in Shetland as in London, and so can be specifically defined. However, the weather has a higher probability of being more severe in Shetland than in London and so this is allowed for.

Considerable simplification is accepted as being necessary to yield a manageable analysis, particularly in terms of drag factors, densities of wet snow accretion, types of accretion etc. when applied to the variable length, multi-span sections of an overhead line. Wherever possible, the simplification process uses factors already in use by overhead line designers with adjustments made in the analysis of overall withstand and reliability factors. In other words, there is a trade-off between a risk of failure and the ability of a line to withstand particular loads.

The load-factored approach

General

The load-factored approach combines a probabilistic analysis of applied loads together with a traditional approach to the ultimate strengths of components.

Load-factored design

Load-factored design is generated by consideration of the following components:

- modelling of weather conditions during wet snow storms
- maps of severe weather areas to enable site-specific design

- modelling of loadings on lines
- modelling of how conductors clash in the wind.

Traditional methods of calculating capability in ultimate terms are used. Factors are calculated which allow scaling of the site-specific site loadings such that failure can be predicted where it has occurred in recent extreme storms, i.e. the models are calibrated with historical event data.

The use of this load-factored design with very traditional structural design enabled the preparation of ENATS 43-40 to set a standard that could be shown by modelling to:

- perform better in the exposed areas which had suffered damage in the recent high profile storms
- allow for more economical design in those areas which had had good service from earlier designs.

ENATS 43-40 represents the first stage in developing national acceptance of site-specific overhead line designs for use in the UK. All lines regardless of site had been required to meet nationally determined design criteria and use of these criteria was observed to have produced lines in some areas which provided high levels of reliability and in others frequent failure. The new approach of 43-40 was developed to address this issue. ENATS 43-40 was developed to demonstrate that these designs were achievable nationally at similar costs to those covered by the current ENATS 43-10 and 43-20 (light and heavy construction for 11 and 33 kV lines).

The approach

The approach was to:

1. assess the most aggressive weather-related parameters at each location and height in the UK from the perspective of wood pole lines
2. model the mechanism by which these weather parameters cause mechanical loads on overhead lines
3. compare the theoretical capability of lines based on these considerations with real historical events and thereby develop withstand and capability factors associated with line design for susceptible areas.

The weather maps of the UK produced by this method relate to the statistical probability of wet snow accretion calibrated on historical experience of lines having spans less than 150 m and conductor diameters below 20 mm. This is important. The work all relates to wet snow accretion and not to rime ice as wet snow blizzards were considered to be the cause of most major failures. An example is shown in Figure 4.1. Note that a separate map is available for each 100 m increase in altitude.

Recent European standards introduced in the UK have implemented weather maps from BS 8100 rather than ENATS 43-40. This subject will be discussed in detail later in this chapter.

Loads and deflections

Storm information was made available from the Meteorological Office from 240 weather stations as half hourly data in many cases for a 14-year period. These data

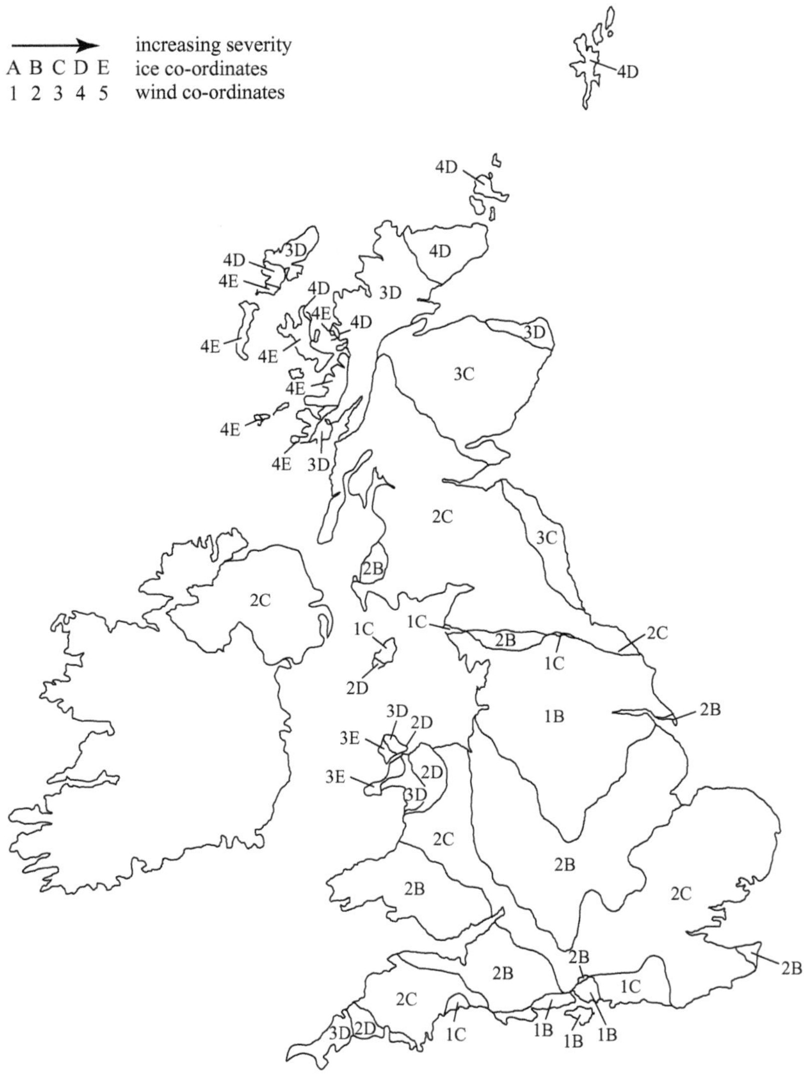

Figure 4.1 ENATS weather map for land 200–300 m (courtesy ENA)

were processed to estimate six-minutely data at each site and at each of six equivalent heights both above and, if appropriate, below the level of the site.

The data were again processed to estimate mean wet snow accretion during each period when accretion was likely at these 240 equivalent weather stations. These data were then analysed to estimate the worst build up of wet snow and of wind pressure likely with a 50-year return period during episodes when wet snow was likely at the height of a wood pole overhead line in a rural environment of reasonably rolling countryside.

A new model was created (based on the above data) for conductor motion and clashing. In such conditions this model was demonstrated to give practical results and to account for failures in previous storms that had not been accounted for until that time.

A new model was created and approved by the Meteorological Office for the mapping of these extreme data and assessing the likely extreme weather-related loads at positions other than those from which data had been gathered. Once again this model appeared to be robust in that exclusion of any data in turn had little effect on the whole spectrum of data representing the country. If the exclusion of some data had affected the model then it would have been too sensitive for general application. It should be noted, however, that no amount of inspection could assess the absolute validity of the model only the relative apparent validity.

Component capability

Earlier approaches had developed an effective database of accepted component capability. Much of this had been the subject of more or less rigorous testing at some stage, yet the statistical data needed for a fully probabilistic approach were not available.

For example, it was known that pole tests had been undertaken to confirm the techniques and assumptions used in the assessment of pole strengths. It was not, however, clear if sufficient tests had been done across a sufficiently representative range of poles to be statistically reliable.

There was a further factor that influenced the decision to accept ultimate failure loads as previously assessed. This was that the new approach had to be acceptable and comprehensible to overhead line engineers of the time who over the years had come to empathise with these methods.

However, this led to certain problems, e.g. in the context of ENATS 43-40 in particular insulator pin capabilities used manufacturers' certified minimum failing loads to represent capability, but it seems this load did not take account of the deflection (under load) of the pin at these and lower loads. Unfortunately, such pins used at heavy pin angle positions deflected alarmingly in normal service and revisions of ENATS 43-40 had to down rate the ultimate failing load of these components accordingly.

Load factors

We are now armed with the weather-related loads, the component capability, traditional conversion from one to the other, methods of mapping and national storm damage reports. It is possible to assess the factors needed to ensure that the consequences of previous storms would be less severe in the future if the new design approach were adopted.

4.2.6 ENATS 43-40 Issue 1

4.2.6.1 General

ENATS 43-40 was therefore developed as an example to show what would be the consequences of adopting the ENATR 111 approach to lines that were in nearly all respects similar to lines that were being erected at the time. Larger conductored lines had previously been erected to one of the heavy designs, the latest at that time being ENATS 43-20. Small conductored lines were erected to one of the light designs with BS 1320 being the subject of the Baldock recommendations and little changed in its various editions up to and including ENATS 43-10.

ENATS 43-40 was accordingly pitched to show that low cost lines, like the light designs, could still continue to be erected in those areas where they had served well in the past. Indeed, it was shown to be reasonable to erect lines with larger conductors to similar designs in these areas. It was also pitched to show the cost of attempting to provide security equivalent to that previously provided for heavy lines in the more exposed areas if small conductors were to continue to be used in these areas.

The layout of ENATS 43-40 reflected the probabilistic base of the design by providing a benchmark approach.

4.2.6.2 The full specification

The specification used structures very similar to those in use in current standards at the time but incorporated minor modifications to improve structural efficiency. Additionally, some supports have been specifically modified to address the principle of failure containment.

To use the specification in practice the engineer uses maps to decide the weather parameters that are likely to represent the site. This reflects the height above sea level of the site, correcting from local experience for any relevant topography that is likely to have an adverse effect. Knowing the conductor that is required, the engineer then converts these extreme weather conditions into equivalent conductor loads using the tables. A check is made on the maximum spans that may be used both in terms of conductor strength and propensity to clash during storm conditions. Propensity to clash is also a function of conductor spacing and so an awareness of the supports likely to be required is needed, although this tends to be a slightly iterative process so nothing needs to be firm and fast at this stage.

The engineer is then aware of the:

- worst design wind and ice expected at the site incorporating a probabilistic assessment
- equivalent ultimate conductor loadings which should be catered for incorporating the load factors and reliability factors that are characteristic of this specification
- maximum spans, which should be used for the chosen conductor, conductor spacings and clashing performance.

Based on the ultimate conductor loadings and knowledge of the support arrangements that provide the required conductor clearances, the engineer can now use the tables provided to determine the support requirements in terms of required strength.

A further iteration is required to determine the failure containment principle. In essence, there is a requirement to provide failure containment measures wherever the designed line is in an area where, in extreme conditions, an unacceptable risk of conductor failure or clashing exists.

Where relatively strong conductors are used with wide conductor spacing, this requirement is considered to be inherently met. Where specific failure containment measures are required the requirement is considered to be met provided specific structure types are used with a specified frequency throughout the line. Again, the concept is that the line shall withstand without failure all likely extreme conditions and shall offer a level of resistance to cascade failure. This occurs when the failure of one component (usually a broken conductor) causes a pole breakage, which then overloads the next pole etc., and a series of poles collapse in a domino fashion until typically a section pole is reached which is strong enough to withstand further collapse.

4.2.6.3 Variants of the specification

It was envisaged at the time that ENATS 43-40 was prepared that other specifications using the ENATR 111 methodology would be prepared as mentioned earlier. In particular, it was envisaged that designs with both higher and lower resistance to the likely extreme weather conditions identified for each location would be needed to meet the requirements of each company.

There are situations where it may be desirable to re-build existing lines to ENATS 43-40 standards, but where applying it strictly is impractical. In such cases ENATR 111 provides an umbrella of protection for the engineer who chooses to develop his own ENATS 43-40 variants.

As has been explained earlier in this chapter, deterministic design is based on well-defined specific loads and strengths, and sets margins of safety between capability and stress. Probabilistic design accepts that all real components have a strength distribution rather than a specific single value. This is particularly significant for natural components such as wood poles. ENATS 43-40 is based on a combination or semi-probabilistic design that uses a specific deterministic design strength subject to a probabilistic range of load conditions. ENATS 43-40 was borne out of a technical specification ENATR 111 that was produced from historical knowledge and meteorological data. The ENATR 111 methodology was demonstrated in the development of the only national specification designed at that time in anticipation of the commencement of the 1988 regulations. This specification, ENATS 43-40, was approved before the regulations took effect.

It is actually notable that ENATS 43-40 was intended only to demonstrate the practicality of using the ENATR 111 methodology. Thereafter it was intended that alternative specifications to meet the particular needs of individual companies with both higher and lower needs in terms of reliability and, conversely, cost would be drawn up.

4.2.6.4 Design loads

Wind and ice loads

In ENATR 111 the weather load combines the factors of ice and wind pressure. The component load takes into account the gust factor (normally taken as 1.5 $\times$ wind speed) in assessing the overall load. The conductor loads can be split into horizontal and vertical components.

The vertical component is the mechanical result of the conductor weight and the ice weight and is known as the maximum conductor weight (MCW). The horizontal load is made up of the wind pressure acting on the ice accretion envelope. This is known as the maximum conductor pressure (MCP).

The combination of these forces is the resultant, i.e. the maximum conductor resultant (MCR), that may occur at any particular point along the component load line.

In the case of angle poles the effect of conductor tension can be significant. The maximum conductor tension (MCT) is defined as the maximum conductor tension at 0 °C at the MCR loading.

Further considerations of downpull etc. are also required but will not be considered here. Briefly, because in flat country the conductors will sag between the poles, they will be applying a downward force or downpull on the poles (trying to force them deeper into the ground – a problem in peat bogs). Normally this downpull angle is assumed to be 1:10. However, if the line crosses a valley and there is a pole in the dip, the conductors may go up from the pole, applying an upward force to the pole (especially in cold weather as the conductors contract). This is known as uplift and the conductors could be trying to pull the poles out of the ground on a badly designed line.

Section 4.2.8 looks at the deterministic snow/ice loads that are part of BS EN 50341/BS EN 50423 that are now adopted for new lines in the UK. This is a simpler version than in ENATS 43-40 Issue 1 and is similar to that used in ENATS 43-20.

Load-factored design

In preparing ENATR 111 the industry had supported a load-factored design approach for wood pole lines in preference to a full probabilistic design base similar to that applied in the then new steel tower design base to reflect the wider coefficients of variability associated with the materials and loading cases in wood pole design.

The load-factored approach used a more or less rigorous probabilistic analysis of applied loads and of anticipated conductor deflections together with a more or less traditional approach to the ultimate strengths of components and a geometric, partial-factored approach to set reliability levels.

Weather zones

It was appreciated that the weather load would not be the same for all areas. Meteorological data was used to generate wind ordinates (1 to 6) and wet snow ordinates (A to E) for different regions or zones of the UK (Table 4.2). As can be seen, the wind ordinate increases from 190 N/mm^2 wind pressure in steps of 190 N/mm^2 and the snow ordinate from 5 mm of radial ice in 5 mm steps. In practice, the wind ordinate, 6, only occurs for land at >400 m in northern and western Scotland and rarely in places where overhead lines are built. In these areas there is in fact no land at this height and so effectively the wind ordinate ranges from 1 to 5. Most of the UK where wood pole lines are constructed is covered by a 2B or 2C weather zone (380 N/mm^2 wind pressure and 10–15 mm of radial ice load).

Table 4.2 Wind/ice load areas

Wind ordinate	1	2	3	4	5	6
Wind pressure (N/mm^2)	190	380	570	760	950	1140
Ice ordinate	A	B	C	D	E	
Radial ice thickness (mm)	5	10	15	20	25	

Software design

The original design tables etc. published in ENATS 43-40 Issue 1 could be replicated and amended for differing conductor arrangements using proprietary software in MS-DOS and SuperCalc formats as issued with that specification. Details of these packages can be obtained from the Energy Networks Association, although the SuperCalc program used for the development of the spreadsheet design folders is now not readily available.

4.2.7 From 1988 to date

Forty years after the industry was nationalised the whole process was reversed. Legislation for privatisation was enacted in the form of The Electricity Act 1989 and, in common with previous acts, its contents were primarily concerned with organisational, legal and fiscal matters.

Similarly, a new set of regulations was introduced – The Electricity Supply Regulations 1988. These regulations represented a significant change in the philosophy of electricity regulations. Whereas previous regulations had given specific design parameters, the 1988 regulations gave very few, in fact, within the actual design parameters, only the Schedule of Ground Clearances survived. Henceforth, the responsibility for selecting design parameters would rest squarely with the electricity supplier. This new philosophy is encapsulated in Part V, Section 17 of the regulations which states:

> All supplier's works shall be sufficient for the purposes for, and the circumstances in which they are used and so constructed, installed, protected (both electrically and mechanically), used and maintained as to prevent danger or interruption of supply so far as is reasonably practicable.

In 1988, therefore, a position was reached where the design engineer, without assistance from a statutory list, must select/devise parameters and components to construct a line which is 'fit for the purpose' taking into account all the technical, safety, geographical, climatic, amenity and environmental conditions that might be encountered.

To the standards engineer and the line design engineer this new approach was both an added responsibility and an opportunity for innovation. However, when privatisation arrived, the Electricity Council's system of working parties more or less ceased for a period of some years. This raised the likelihood that each DNO with its newly found freedom to innovate would begin to develop its own standards. The spectre of declining standards across the UK with some 12 versions of the same basic design could bring about consequential manufacturing and commercial difficulties.

4.2.7.1 The Electricity Safety, Quality and Continuity Regulations, 2002

These regulations (known as ESQCR 2002) were published in October 2002 and came into force from 31 January 2003. They cover:

a protection and earthing
b sub-stations
c underground cables
d overhead lines
e generation
f supplies to other networks
g design schedules.

Here, this chapter covers only items *d* and *g*. The regulations again refer to either bare or insulated wire, and therefore continue the previous regulations in treating covered conductors as bare wire, unless they have sufficient covering to be regarded as insulated. The regulations also put the responsibility of safety squarely with the owner of the overhead line. The term 'ordinarily accessible' is used. This means that the OHL should not be reachable by hand 'if any scaffolding, ladder or other conducting object was erected or placed in, against or near to a building or structure'. If someone wishes to erect a building so that it may allow an OHL to be ordinarily accessible then all they have to do is give reasonable notice to the DNO. These definitions leave the industry open to many problems concerning permanent or temporary structures (e.g. tents) built under or near lines.

The section on anticlimbing devices (ACDs) is also woolly and open to many interpretations. It states that every support shall 'if circumstances reasonably require' be fitted with ACDs to prevent 'so far as is reasonably practical' any unauthorised person reaching a 'position of danger'.

Schedule 2 of item *g* specifies the minimum height above ground of overhead lines (see Table 2.1).

4.2.7.2 Other standards

For many years the UK standards listed at the end of this chapter (section 4.4) have been the design bedrock of the UK overhead line networks. Nowadays the hierarchy of standards is on an international scale and standards should be selected and used in the following sequence dependent on their availability:

1 international standards
2 European standards
3 British standards
4 Electricity Association technical specifications
5 company standards.

One of the most important recent standards for lines at all distribution voltages is BS EN 50341 (a CENELEC standard) and its sister draft publication

BS EN 50423. Both these standards have a common UK national normative annex (NNA) BS EN 50423-3-9.

4.2.8 BS EN 50341 and BS EN 50423

BS EN 50341 refers to overhead electrical lines exceeding AC 45 kV and BS EN 50423 refers to overhead electrical lines exceeding AC 1 kV up to and including AC 45 kV. Essentially, the UK has to adopt these new standards and in particular their view on wood pole design using either probabilistic methods (called the general approach in these documents) or deterministic methods based on traditional factors of safety (known as the empirical approach). It is this latter empirical approach that the UK BSI standards committee has declared will be adopted for all wood pole line designs.

The general (or probabilistic) approach was seen as far too onerous for the UK in respect of wood pole designs due to the following reasons:

1 The characteristic fibre stress (the fibre stress of the bottom 5 per cent of the spread of fibre stress values found in wood poles is effectively taken as the characteristic value in the UK after the use of factors of safety) would be lower than currently used. Typically, the mean fibre stress for Scots Pine as purchased by the UK DNOs is 53.4 N/m^2, whereas around 5 per cent of these actually have a fibre stress of below 21.4 N/m^2 due to the normal range of properties of a natural product. At the time of writing, no supplier in Europe has published an appropriate characteristic stress that should be adopted for a probabilistic design approach for each potential pole species.
2 Ice load is very large for no wind conditions, so very high line tensions need to be considered.
3 Vertical loads for intermediate poles must be considered. Overturning loads, only, have historically been considered for these structures.
4 A ten per cent deflection limit was considered very onerous for intermediate poles, especially for the taller poles.

To take account of UK experience in wood pole line designs, a NNA for the UK was agreed to be associated with the main body text of BS EN 50341/50423. Known as BS EN 50423-3-9, this part of the standard details the specific empirical design approach to be employed for wood pole lines and allows the UK to specify loading scenarios and factors of safety very similar to those included in ENATS 43-20. It should be noted that the load factored design approach employed in ENATS 43-40 Issue 1 has now been superseded by this new design approach, which has now been incorporated within Issue 2 of that technical specification.

There now follows a brief description of BS EN 50341/50423 and the UK NNA BS EN 50423-3-9 as it applies to UK wood pole OHL at voltages up to and including 132 kV (trident design). It does not apply to LV lines ($<$1000 V AC). The standard need only apply to new OHL and not to the maintenance, re-conductoring, tee-offs, extensions or diversions to existing lines.

Table 4.3 Wind pressures and aerodynamic drag factors

Load condition	Wind pressure (N/m^2)		Aerodynamic drag factors	
	q_x	q_c	C_x	C_c
High wind (no ice)	1740	1740	0.8	1.0
Combined wind and ice (normal altitude)	380	380	1.0	1.0
Combined wind and ice (high altitude)	570	570	1.0	1.0
Wind only (no ice)	0	760	–	1.0
Security (broken wire)	380	380	1.0	1.0

For the leeward (shielded) pole, a shielding factor of 0.5 shall be assumed.

Wind loads

The wind pressures and aerodynamic drag factors are given in Table 4.3. The span factor, G, is assumed to be 1.0 for wind spans $\leq$200 m and $(0.75L + 30)/L$ for wind span lengths $(L) >$200 m.

The specification also defines high and normal altitude as:

normal altitude	for all GB except Scotland for site altitudes 300 m in Scotland <200 m
high altitude	all other locations up to 500 m.

The standard does not apply to lines on land above 500 m in the UK.

Wind and ice loads

For conductors not exceeding 35 mm^2 copper (60 mm^2 aluminium-based) conductors, the wind only case in Table 4.3 may be used. For all other lines the radial ice thicknesses are:

normal altitude	9.5 mm
high altitude	12.5 mm

The ice density is taken as 913 kg/m^3, although other ice densities of 500 kg/m^3 (for rime ice) and 850 kg/m^3 (for wet snow) may be specified. The minimum temperature is as traditionally used in the UK, i.e. $-5.6\,^{\circ}C$.

Conductor loads

Conductor load cases are specified in Table 4.4.

Factors of safety

EN 50423-3-9 includes partial factors which are essentially the UK's factors of safety (FOS). A list of the FOS is given in Tables 4.5 and 4.6.

It is important to note that Table 4.6 was derived to simplify the effect of vertical loads that continue to be ignored for intermediate wood pole structures. OHL components also have partial factors applied, known as gamma-m factors, as given in Table 4.7.

Table 4.4 Conductor load cases

Load cases	Temperature (°C)	Load condition	Notes
1	−5.6	high wind (no ice)	as detailed in the project specification
2	−5.6	combined wind and ice (normal altitude)	normal altitude: conductor >35 mm^2 copper
3	−5.6	combined wind and ice (high altitude)	high altitude: conductor >35 mm^2 copper
4	−5.6	wind only (all altitudes – no ice)	conductor up to 35 mm^2 copper equivalent area
5	−5.6	security (broken wire)	as detailed in the project specification
6	−5.6	construction and maintenance	see project specification

Table 4.5 Partial factors for actions, ultimate limit state

Action (load)	Partial factor
Normal load cases – variable actions	
Climatic loads and conductor tension	
high wind (load case 1)	1.1
combined wind and ice (load cases 2 and 3)	2.5[2]
wind only (load case 4)	2.5[1,2]
Permanent actions	
Self weight	
high wind (load case 1)	1.1
combined wind and ice (load cases 2 and 3)	2.5[2]
wind only (load case 4)	2.5[1,2]
static cantilever loads (all load cases)	1.0
Exceptional load cases – security (broken wire) loads (load case 5)	1.3
Construction and maintenance (load case 6)	1.5 on static loads 2.0 on dynamic loads

[1] For timber pole supports, wind on the pole is ignored.
[2] Higher partial factors may be specified in the project specification, particularly for intermediate poles. See also Table 4.6.

Table 4.6 Partial factors for actions, intermediate pole declination

Action (load)	FOS
Declination gradient – climatic loads	
Level – 1:10 (load cases 2, 3 and 4)	2.5
>1:10–1:7.5 (load cases 2, 3 and 4)	3.0
>1:7.5–1:5 (load cases 2, 3 and 4)	3.5

4.2.9 Clashing

Bare phase conductors need to be spaced to avoid mechanical and electrical damage due to clashing. Covered conductors do not, as their sheaths are easily capable of several million clashes without damage. In the case of bare conductors, for wood pole lines at normal altitudes, the minimum recommended phase separation is defined by weather zone 2B, and, for lines at high altitude, the minimum recommended phase separation is defined by weather zone 3C. Greater phase separations may be required due to the effect of funnelling or for altitudes greater than 500 m.

ENATR 111 defines the weather zone applicable to an area. This is where the likely mean wind pressure and absolute maximum ice accretion thickness may be described by a numeral and letter, respectively. The wind co-ordinate is described in 190 N/m^2 increments, and the ice co-ordinate is measured in 10 mm diametric thickness increments for each letter increment (A = 10 mm, B = 20 mm etc.).

The gust and lull wind pressures are 1.832 and 0.546 times the mean wind pressure, respectively. The minimum spacing to avoid conductor clash is based on the worst combination of wind and ice.

Maps of weather zones are shown in 100 m increments of elevation above mean sea level in ENATR 111. However, as far as the NNA is concerned, 2B represents a wind pressure of 380 N/m^2 with 10 mm of radial ice load, whereas 3C represents 570 N/m^2 and 15 mm of radial ice.

4.2.10 Wood poles

As has been stated, wood poles will normally be fabricated using *Pinus Sylvestris* (Scots Pine) taken from a population whose southern boundary lies at a latitude of 60° north. The poles are accepted to have the following characteristic values:

mean bending strength (modulus of rupture)	53.3 N/mm^2
mean modulus of elasticity	10 054 kN/mm^2

Where differing species are to be used, this will be defined in the project specification.

Table 4.7 Partial strength factors for overhead line components for the empirical approach

Component	Material property	γ_m
Steel members (grade S275) used as ancillaries on wood poles	resistance of cross sections and buckling of sections (based on yield strength)	0.64[3]
	resistance of bolted connections (based on ultimate tensile strength):	
	• shear	1.33
	• tension	1.0
	• bearing	2.0
	resistance of welded connections (based on yield strength of parent steel)	0.64[3]
Timber poles	resistance of body of pole, cross-section, elements and bolted connections (based on mean ultimate strength)	1.0 min[1]
Guyed structures	resistance of guys (based on nominal failing load)	1.0 min
Foundations	refer to project specification	
Conductor	resistance of conductors (based on nominal breaking load)[4]:	
	combined wind and ice	0.8
	wind only	1.0
Tension, suspension, pin and post insulator sets[2]	all string components (based on nominal failing load)	1.0 min

[1] Based on the application of stresses as defined in Clause 7.5.5/GB.4 of this NNA. For wood pole intermediate unguyed supports, the effects of the vertical loading are ignored unless specified in the project specification.
[2] The coefficient applies only to ceramic (glass and porcelain) insulators: where non-ceramic insulators are to be used the coefficient will be defined in the project specification.
[3] The appropriate γ_m factor has been determined based on the ratio of yield strength to ultimate tensile strength assuming grade S275 steel to EN 10025. For other steel grades, use same ratio to provide γ_m value.
[4] The nominal breaking load of conductors is a client defined percentage of the rated strength of the conductor as given in the appropriate standard, e.g. EN 50182.

4.2.11 ENATS 43-40 Issue 2 (2004)

This now follows the wind/ice loads as specified under BS EN 50423-3-9 except that it is now allowed to use densities of 913 kg/m^3 for glaze ice, 850 kg/m^3 for wet snow and 510 kg/m^3 for rime ice. Where specific data are unavailable, the standard recommends that the density value for glaze ice should be used. This specification also includes an MS Excel version of the original 43-40 software, but now modified to comply with BS EN 50423-3-9 (UK NNAs) and retains the original ENATS 43-40 Issue 1 clashing requirements.

4.3 Summary

In essence, the electricity industry supported ENATR 111 and presented it to the Department of Energy representatives (who had been members of the Baldock Enquiry) as a half-way house between deterministic and probabilistic principles for future wood pole designs. Load-factored design has been used in the introduction of the site-specific design methodology that was established to meet the requirements of the 1988 electricity regulations for the design of wood pole overhead lines. This approach was then endorsed with the 'deemed to comply' epithet such that the industry contended that lines designed in this way met the 'fitness for purpose' clause of the 1988 regulations. The new ESQCR 2002 regulations still leave the onus on the distributor to ensure that lines are safe, reliable and in all ways 'fit for purpose' as far as 'reasonably practical'. The recent CENELEC standards BS EN 50341 and BS EN 50423 have brought in deterministic and probabilistic designs for new lines under the UK NNA BS EN 50423-3-9, although in respect of wood pole lines only the deterministic approach is declared.

4.4 Relevant OHL standards

ENATS 43-120	Fittings for covered conductors from 1 to 33 kV
ENATS 43-121	Specification for single circuit overhead lines of compact covered construction on wood poles for use at high voltages up to and including 33 kV
ENATS 43-122	XLPE covered conductors (11 kV and 33 kV overhead lines)
BS EN 50341	Overhead electrical lines exceeding AC 45 kV
ENATS 43-40 (2004)	Overhead line conductors up to 33 kV
BS EN 50423	Overhead electrical lines exceeding AC 1 kV up to and including AC 45 kV
BS EN 50423-3-9	UK NNA for overhead electrical lines exceeding AC 1 kV
ESQCR	Electricity Supply Quality and Continuity Regulations 2002
ESI Regulations, 1970	The Electricity (overhead lines) Regulations 1970
ESI Regulations, 1988	The Electricity Supply Regulations 1988 (as amended)
BS 1990 part 1	Wood poles for overhead lines (power and telecommunication lines)
ENATS 43-10	11 kV single circuit overhead lines of light duty construction on wood poles
ENATS 43-20	11 kV and 33 kV single circuit overhead lines of heavy duty construction on wood poles
ENATS 43-40	High-voltage single circuit overhead lines on wood poles

ENATS 43-88	Selection and treatment of wood poles and associated timber for overhead lines
ENATS 43-90	Anticlimbing devices for HV lines up to and including 400 kV
ENATS 43-91	Stay strands and stay fittings for overhead lines
ENATS 43-93	Line insulators
ENATS 43-95	Steelwork for overhead lines
ENATS 43-96	Fasteners and washers for wood pole overhead lines
ENATR 111	Report of the high-voltage single overhead lines on wood poles

Chapter 5
Overhead line design

5.1 Historical review

Throughout the twentieth century electricity became arguably the most important commodity provided to the home and industry and now interacts with almost every aspect of our lifestyle. Distribution overhead line designs supplying electricity up to 33 kV, unlike electrical goods, have not necessarily changed all that much, yet there have been significant developments in design criteria that should be noted. Overhead line reliability and safety have been heavily criticised in recent years and so there is a need to address the issues of future overhead line design. This chapter looks at some of the fundamental steps that are adopted in developing a new overhead line wood pole design for the UK environment together with some historical references and the effects of new European legislation.

5.2 Background

A detailed review of design standards is given in chapter 2 from a regulatory viewpoint. In this section the relevant standards are considered from a design viewpoint.

Regulation 15 of the Electricity Supply Regulations (1937) stated 'overhead lines shall be erected and maintained in accordance with the provisions of any regulations made under the Electricity Supply Act 1882–1936'. The regulations specifically applying at that time were the Overhead Line Regulations of 1931 which, for all lines above 325 V, laid down the following design standards:

Conductors – factor of safety 2.0 on the design conditions of wind pressure of 8 lbs/sq ft acting on 3/8″ radial ice-loaded conductor at 22 °F.
Wood poles – factor of safety of 3.5 with wind pressure of 8 lbs/sq ft acting on 3/8″ radial ice-loaded conductor at 22 °F.

The 1947 revision of the overhead line regulations, E1.C.53 (1947), did not change the design standard but stated in Regulation 21 that 'lines may be erected in accordance with BS 1320 dated 1946'. This specification relates to conductor sizes smaller than 35 mm^2 copper equivalent.

The Overhead Line Regulations (1970) laid down the same design parameters as BS 1320 for lines with conductors up to 35 mm^2 (0.054 sq in) cross section apart from making reference to 3.5 factor of safety for home grown poles. For lines above 35 mm^2 copper equivalent the 1970 regulations laid down the same parameters as E1.C.53 (1947), i.e. wind and ice loading, apart from relaxing the factor of safety on poles to 2.5.

ENATS 43-10 (1974) was issued as a new design standard for 11 kV lines of light construction to meet with the requirements of the OHL Regulations 1970. This was a metric standard with a straight conversion of a wind pressure of 16 lbs/sq ft to 760 N/mm^2 and 22 °F to −5.6 °C.

ENATS 43-10 also includes an additional design standard for lines in severe environments which increases the factor of safety for supports from 2.5 to 3.5, reduces recommended span length by 20 per cent and limits the conductor size to 35 mm^2 copper equivalent.

ENATS 43-20 (1979) was issued as a new design standard for 11 kV and 33 kV lines of heavy construction and covered aluminium conductor steel reinforced (ACSR) conductors only. This standard re-iterates the design conditions laid down in the 1970 regulations and again caters for lines in severe environments by increasing the support factor of safety and reducing recommended span lengths. It can be seen that although there have been a number of documents issued in the past the mechanical load figures have not necessarily changed, just the factors of safety. Later in this chapter we will be coming back to this standard as a way to keep in line with new European standards.

In the early 1980s, due to severe storms experienced in the south of England, a joint panel of enquiry was set up involving both the government and representatives of the electricity supply industry to review technical standards on overhead lines.

A review of current UK designs and a comparison of these to European equivalent designs were undertaken. UK and European meteorological data were examined and advice taken from independent OHL consultants.

The report found that the most serious and common factor that contributed to faults was mechanical overload failure of the conductors on 11 kV duty lines to BS 1320. In addition, the long span lengths adopted increased the likelihood of conductor clashing. Although improvements had been made with the introduction of ENATS 43-10 this design was also found to be inadequate in the severe weather conditions experienced with the data available.

The main recommendation submitted by the panel of enquiry, which was subsequently adopted, was that the electricity supply regulations be revised to ensure that statutory requirements are more closely related to the weather conditions actually experienced in an area.

In 1988, the electricity supply industry produced a national design specification for overhead lines that was prepared and issued specifically to satisfy the requirements of the new regulations. The specification (EA technical specification 43-40) used a new approach to overhead line design. The approach taken matches the maximum likely weather-related load at specific locations with the strength capability of the weakest component.

In brief the three main issues for overhead line design using the models described are:

1 maximum conductor weight (MCW)
2 maximum conductor pressure (MCP)
3 maximum conductor tension (MCT).

5.3 Technical requirements for line design

5.3.1 The conductor

The first component to fail on well-constructed and maintained lines is invariably the conductor, either through overload by ice (strictly wet snow accretion) and/or wind at or about the conductor's nominal breaking load. During high winds it is also possible for conductor clashing to occur, which will cause the conductors to burn at the point of contact thus reducing their overall strength.

It is essential that when looking at design from first principles the overhead line conductor is the most important consideration. The technical data relating to each conductor's characteristics are used to determine the limits of the conductor under loaded conditions and at various temperatures relating to the basic span chosen.

5.3.1.1 Vibration limit

The design limits used are generally based on the maximum vibration limit applicable to the materials used. For distribution overhead lines the wind-induced vibration and oscillation manifests itself in the following modes:

a Galloping – this is a low frequency, large amplitude oscillation generally caused by winds of 5 to 10 m/s. It can occur on very long spans such as river crossings and exposed mountainous terrain with typical amplitude of three metres.

b Aeolian vibration – this type of vibration occurs when a steady wind of fairly low velocity (1 to 16 m/s) flows across cylindrical objects. This causes vortices to be shed on the leeward side creating forces in alternating up and down directions. Vibration is therefore vertical having amplitudes up to the conductor diameter. The frequency of vibration is related to the natural frequency of the conductor and is in the region of 3 to 100 Hz.

Modern thinking on vibration limits has changed in recent years and this will be covered in chapter 7.

5.3.1.2 Maximum conductor tension

The maximum conductor tension is regarded as the tension at −5.6 °C with the maximum applied load in ice and wind. This load must not exceed the safe limit in relation to the maximum working tension of the conductor and so factors of safety (FOS) of 2 and 2.5 have been used in the past.

The loads from the pressure of the wind on the conductor (MCP) and the weight of the conductor material together with the ice accretion (MCW) have a resultant load, which is calculated in addition to the other conductor parameters to provide a maximum conductor tension. It is this figure that must not exceed the requirement of the previous paragraph at −5.6 °C. These figures are then used to calculate whether the cross-arm configuration is capable of withstanding the forces applied.

5.3.2 Cross-arm design parameters

The cross-arm configuration strength to withstand imposed loads due to conductor stresses is evaluated after considering the following parameters:

- electrically sound
- weight
- reliability/durability
- flexibility (utilisation of standard components).

The most common cross-arm configuration currently used in the UK is the standard intermediate horizontal type typical of BS 1320 and ENATS 43-10.

The ENATS 43-20 intermediate cross-arm is wider (at an overall length of 2.5 m) and weighs one third more (42 kg) than the earlier ENATS 43-10 design (28 kg).

Making further reference to the BS 1320/ENATS 43-10 designs, these overhead line specifications standardised the cross-arm design for conductors up to and including 32 mm^2 HDCu or 50 mm^2 ACSR and for voltages up to and including 11 kV. Overhead lines with larger conductor sizes have adopted the wider cross-arm referred to in ENATS 43-20 and ENATS 43-40.

Cross-arms to the specifications mentioned to date have been designed with a factor of safety of 2.5 applied on the ultimate strength of the material, and it has been noted that in general terms these cross-arms have performed satisfactorily in service for many years.

5.4 Designing horizontal cross-arms for single supports

Cross-arms for single supports can generally be treated as two cantilevers fixed at the support. The cross-arm at the support is therefore subjected to the following bending moments (BM):

1. BM due to the weight of a span of ice-coated conductor acting in a vertical direction at a prescribed distance from the pole centre.
2. BM due to span of wind-loaded conductors acting at a distance above the centre of the cross-arm (assuming pin or post type insulators).
3. BM due to alterations of profile. This may be either a positive or negative component of line tension, depending upon whether the profile imposes a downpull or an uplift on the conductors.

Further details regarding cross-arm design are given in chapter 6.

5.5 Vertical and strut loadings

Where section angle or terminal structures are considered, then the forces acting on the pole top are far greater. The horizontal loading being applied to the top of the structure must therefore be counteracted in some form. This is generally achieved by a stay wire formation.

Due to the combination of both horizontal loading and stay strut loading, a vertical strut load is applied to the structure. The forces acting on the pole are therefore dependent on the stay formation and the slope or angle it falls from the pole. The more acute the angle of stay slope the greater the vertical loadings imposed on the structure. It is therefore important to try and achieve the greatest stay angle possible and this is typically 45° where the resultant line tension will be equal to the strut load due to conductor load.

It is also important to account for the following additional loads when calculating the absolute load on the pole:

1 the weight of the conductors for the span length (three conductors × windspan × weight/m)
2 the weight of the pole top fittings (insulators and cross-arm steel work)
3 the additional loadings if any for downpull on one or either side of the structure in question.

Once all of these issues have been considered it is then possible to consider matching a structure to the loads calculated. The crippling forces obviously must fall below the maximum permissible strut load that the wood pole structure can accommodate.

5.6 Support design

BS 1990 provides details of typical sizes for Scots Pine (*Pinus Sylvestris*), the preferred wood pole support of the electricity supply companies in northern Europe. This specification, however, merely provides guidance in relation to the typical strut strengths of the typical poles provided relative to grade and diameter.

Reference tables are usually provided for the line design engineer to determine which pole should be chosen. If the maximum pole strength cannot be accommodated in the spreadsheet then it may be necessary to consider using a different support type such as an 'H' pole configuration. The crippling load imposed on one structure can now be shared accordingly between the two poles. This may, however, not be an equal share depending on how the stays are arranged.

5.7 Windspan and foundation

Three factors can effect whether a pole is suitable for use, one we have just discussed being its crippling load, the two others are its windspan capability and its foundation strength.

5.7.1 Windspan

There is a maximum strength in a pole based on its ability to withstand horizontal loads. The wood pole can only withstand a degree of pressure applied to the top of the structure in a horizontal plane based on the span of conductor and pressure applied to the conductor length and area. The values given for typical windspans are a function of its:

- modulus of elasticity
- diameter at the groundline
- diameter at the point of application of the load
- distance from the groundline to the point of application of the load.

5.7.2 Foundation

The foundation capabilities are based on the horizontal forces applied at the top of the structure and the foundation's ability to withstand this force. The foundation's resistance to withstand this applied force is a function of its stability (condition of the soil) and the resistance area of the pole and any associated blocks below the ground line against the soil. The resistance can be improved by changing the ground conditions (imported backfill or concrete) or increasing the area of resistance with the addition of baulks or increased depth by auguring.

5.8 UK line design for the future

5.8.1 General

For existing lines and any modifications or extensions to them, the appropriate line design specification as described in chapter 4 can be used. In terms of new lines, BS EN 50423 and BS EN 50341 will progressively be introduced and employed. In essence, the design approach in these standards is close to reverting to the deterministic ENATS 43-20 situation for lines at normal altitude and then using the higher loads at higher altitudes. This chapter is not intended to be a line design specification as this is beyond the scope of this book, however, the basics of line design on the basis of the new deterministic approach will now be given.

5.8.2 Deterministic design

5.8.2.1 Conductors

Factors of safety are as defined in the previous chapter (Tables 4.5–4.7). Conductors have self-damping characteristics that reduce the amplitude of damaging Aeolian vibrations. Self-damping increases with conductor size but decreases as strands become locked tighter together with increasing tension or compaction. This is one reason why compacted covered conductors cannot be strung at the same tension as bare conductors. Experience has shown that the self-damping of 50 mm^2 aluminium alloy covered conductor requires a maximum everyday design stress (EDS) of 28 N/mm^2

(at 5 °C) to reduce Aeolian vibrations to an acceptable level in open terrain. However, new thinking on Aeolian vibration has come down in favour of using a factor based on the conductor tension divided by the conductor weight/unit length rather than just tension alone. When allowance is made for terrain factors and the use of dampers, the tension levels allowed could be quite different from those based purely on a percentage of the ultimate tensile strength. The EDS can thus be increased where lines are located in hilly/wooded terrain or where vibration dampers are installed.

5.8.2.2 Other components

All other components have to obey the FOS approach defined in the previous section. Wood poles use the forces calculated from wind span and basic/maximum spans as defined by the project specification for the altitude and conductor size used.

5.8.2.3 Design calculations

The software circulated with ENATS 43-40 Issue 2 and ENATS 43-121 can be used for line calculating loads and the suitability of components to resist those loads. This has avoided the need for the line designer to undertake a significant number of hand calculations.

Chapter 6

Mechanical design of poles, cross-arms and foundations

6.1 Scope

Today, there are many software design packages that will allow the OHL design engineer to do all the necessary calculations at the touch of a few buttons. However, it is necessary to appreciate the engineering behind any software, as it is essential to understand what is happening and to feed in the engineering knowledge. This chapter deals with the mechanical aspects of pole and cross-arm strengths and foundation capabilities. The next chapter will deal with the mechanical side of conductors – the tensions in them that cause stresses on the poles and the sag that is relevant for maintaining line clearances in most weather conditions.

In addition, this chapter covers the ground conditions that will affect the choice of pole foundation requirements and describes the importance of stay wires in line design.

6.2 Mechanical design

There is normally a considerable difference between the requirements for a line in the south of England and a line built in the Shetlands. Unfortunately, in 1987, New Year 1999 and 2000 and again in October 2002, the UK experienced the sort of conditions normally only found in the Shetlands and the devastation focused minds once more on line design. This chapter is therefore intended to give you a small insight into the first principles and methods used to produce a line design and some of the problems encountered in this work.

6.2.1 Foundations

The basis of all good construction is the foundation. This has to be able to support the pole under all circumstances within the design parameters of the line.

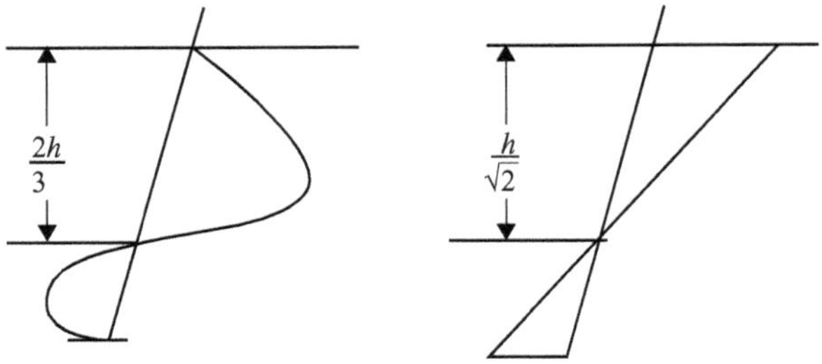

Figure 6.1 Formulae for stress distribution on the pole

6.2.1.1 Soil characteristics

In order to work out what forces the foundation must withstand we must know something about the resistance or holding strength of the soil that the pole is to be planted in. If a tower is to be erected, it is normal for a geotechnical survey to be carried out and the actual value of this strength determined. This is not economically viable for wood pole lines and so some typical values of resistivity are used when designing a foundation.

It is known that a pole will pivot about some point below ground level. There are two formulae, one representing the parabolic form of stress distribution with the fulcrum point taken at $(2/3)h$ from ground level:

$$M_g = \frac{kDh^3}{12}\ \mathrm{Nm} \tag{6.1}$$

where D is the average diameter of pole below ground level in m, h is the depth of planting in m, k is the maximum rupturing intensity in N/m^2/m depth and M_g is the moment of resistance of soil[1]/ground in Nm.

The formula representing the straight-line form of stress distribution where the fulcrum is taken as $h/\sqrt{2}$ (Figure 6.1):

$$M_g = \frac{kDh^3}{10}\ \mathrm{Nm} \tag{6.2}$$

The straight-line formula (6.2) is the one used by most engineers. It is assumed that the intensity of stress is directly proportional to the depth. Soil conditions vary depending on the lateral capacity to support load, and in the following example, average soil has a factor k of 314 175 N/m^2/m (2000 lb/ft^2/ft).

6.2.1.2 Depth of foundation

Expression (6.2) can be used to determine whether the depth of foundation of a wood pole is adequate. In the recent storms, let us say the wind force on the conductors attached to this pole was 5 kN and the wind load on the pole was 2 kN. From

[1] The moment of resistance of the soil is the capability of that soil to resist the overturning moment of the wood pole.

equation (6.2), the moment of resistance of the pole is 55 kNm (assuming a 12 m pole, 0.3 m diameter planted 1.8 m deep).

The conductor wind load is taken to act over the distance from the average mounting height to the fulcrum point, which has been stated as the planting depth/$\sqrt{2}$ from the ground point. The moment of this force is thus this distance plus the ground to conductor level distance multiplied by 5 kN. If the conductor level were say, 200 mm above pole top, the bending moment on the conductors would be 58.4 kNm. Add this to the wind force on the pole from half-pole height above ground to foundation fulcrum, and a total force of over 71 kNm is calculated. In this case, therefore, the wind will cause the foundations to fail and the wood pole structure would be uprooted. The pole could be just planted deeper and further calculation would be needed to assess this suitability. Alternatively, one or more baulks below ground is often considered or, in particularly poor ground areas, the use of stay wires (wind stays) may be considered. In this latter case, these are normally placed square to the line and are not designed according to the normal forces to be experienced due to conductor tension etc.

6.2.1.3 Use of baulks

A baulk is a piece of wood something like a stubby railway sleeper. The standard size of a baulk is $1300 \times 250 \times 125$ mm. Baulks are used to support wood poles by increasing the surface area that is available for ground resistance to overturning or sinking. They are mounted horizontally and, if only one baulk is fitted, it is placed 500 mm below ground level. This is a compromise depth between maximum mechanical advantage and the minimum cover to allow farmers to plough over the baulk.

the area of baulk to resist overturning is $1.3 \times 0.25 = 0.325\,\text{m}^2$
the area of the pole covered by the baulk is $0.3 \times 0.25 = 0.075\,\text{m}^2$
thus the area of the baulk available for reinforcement (i.e. the extra surface area gained by using the baulk assuming the average pole diameter below ground) is $0.325 - 0.075 = 0.25\,\text{m}^2$
hence, additional resistance to overturning is $314.175 \times 0.25 \times 0.5 \times (1.8/\sqrt{2} - 0.5) = 30.3$ kNm
in the example above, the total resistance to overturning is $55+30.3 = 85.3\,\text{kNm} > 71\,\text{kNm}$ okay.

If the pole hole is not to be dug out by hand or by mechanical excavator but is to be augered (essentially using a large earth drill), then fitting of baulks is not easy. Instead, it is necessary to increase the pole planting depth or to increase the pole's effective diameter by backfilling the gap between the pole and the surrounding soil. This can be done with concrete or with a proprietary product that allows the backfill to consolidate chemically.

6.2.1.4 Stays

Another use for the augering equipment is to install stays in very soft ground when it is not possible to obtain the required holding strength using normal excavation

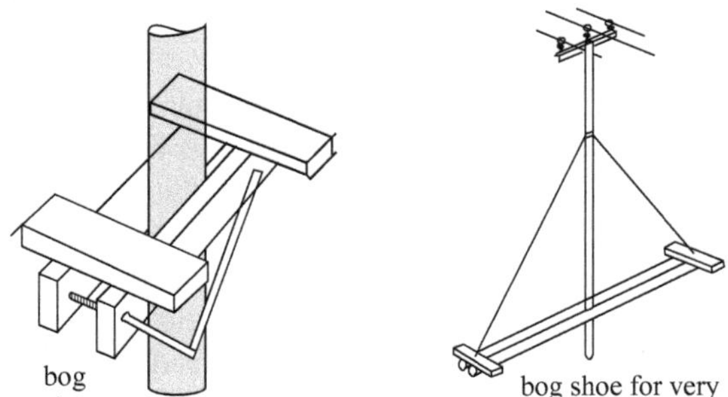

Figure 6.2 Bog shoes

techniques. There are various stay augers on the market that allow extension pieces to be added until the stay eventually reaches a solid layer of ground. This may be as much as thirty feet down in parts of the fen country.

6.2.1.5 Bog shoes

In bad ground it may be necessary to fit a foundation that gives both lateral stability and provides increased bearing capability. This is known as a bog shoe for obvious reasons. The design can vary but, in the worst circumstances, it may consist of two poles laid horizontally across the line and bolted together, the poles are then scarfed to allow the line pole to be inserted at right angles. Wind stays are then fitted from the line pole to the ends of the bog shoe (Figure 6.2). Wind stays are stays mounted normal to the line that hold the pole in position vertically whatever direction the wind comes from. The pole is then literally floated in or on the bog. A good example of this method is the line heading south from the old Spadeadam rocket site on the Northumberland/Cumbria border. Although floating across a peat bog, the poles have not sunk and are still vertical after some fifteen years. This type of foundation is very common in the Hebrides and Shetland.

6.2.2 Wood pole design

6.2.2.1 General

Originally, wood poles were designed as per the 1947 Electricity Commissioners' Regulations, which stipulated that they had to have a factor of safety of 3.5 with a wind pressure of 8 lb/ft^2 (380 N/m^2) and 3/8 in (9.5 mm) radial ice on the conductors at 22 °F (−5.6 °C).

Today, under the 1988 Electricity Supply Regulations, the criterion is that the pole must be 'fit for purpose'. This gives the design engineer much more latitude in his pole design and has led to the environmental approach detailed in ENATS 43-40 already covered in chapter 4.

6.2.2.2 Moments of resistance of wood poles

If it is true to state that the strength of a foundation is dependent on the strength of the ground, then it is also true that it is dependent on the strength of the wood pole as well.

If a pole is subject to a bending moment, then the fibres on one side of the pole are compressed and the fibres on the other side are extended. Compressive and tensile stresses are thereby introduced in the pole producing a moment called the moment of resistance; this is equal and opposite to the bending moment.

The second moment of area or the moment of inertia of a circular section (I) is:

$$I = \frac{\pi D^4}{64} \tag{6.3}$$

The modulus of section (Z) is:

$$Z = \frac{\pi D^3}{32} \tag{6.4}$$

Fibre stress[2] (F) equals:

$$F = \frac{\text{bending moment}}{\text{section modulus}}$$

Maximum resisting moment equals section modulus $\times\ F = (53\,300 \times \pi D^3)/32$, which equates to approximately $5233 D^3$ kNm, where D is the critical diameter of the pole that is either at the groundline or at a point 1.5 times the pole top diameter.

The maximum resisting moment of poles can be obtained from tables. These vary from 28 kNm for a 175 mm diameter pole to 224 kNm for a 350 mm diameter pole. To determine the maximum working resistance of a pole this figure should be divided by the factor of safety adopted for any particular location. This can perhaps be best explained by going through a worked example.

Example

A 12 m intermediate pole is used to support a three-phase 175 mm^2 Lynx ACSR conductor (diameter 19.53 mm), with a span length of 100 m. The pole is planted 1.8 m deep. The wind loading is taken to act at 250 mm above the pole top (the approximate height of the conductors). A factor of safety of 2.5 is normally included in these calculations. Assume the critical diameter occurs at ground level. The aim is to determine what diameter of pole is needed, i.e. medium, stout or extra stout.

First calculate the wind load on the three-phase conductors:

$$= \frac{3 \times 380 \times 100 \times [19 + 19.53]}{1000} = 4393\ \text{N}$$

[2] The ultimate mean fibre stress for wood poles is taken as 53 300 kN/m^2 according to BS EN 50423-3-9.

Wind load on pole assuming average diameter of 250 mm is:

$$\frac{380 \times 250 \times 10.2}{1000} = 969\,\text{N}$$

Wind load on pin/post insulators, assuming projected area of each insulator = $0.15\,\text{m}^2$:

$$380 \times 0.15 \times 3 = 171\,\text{N}$$

Maximum bending moment at ground level equals:

$$\{(4393 \times 10.45) + (969 \times 10.2/2) + (171 \times [10.2 + 0.125])\} \times 1000$$
$$= 52\,614\,325\,\text{Nmm}$$

Maximum allowable fibre stress with a factor of safety of 2.5

$$= 21.32\,\text{N/mm}^2$$

Moment of resistance

$$= f \times \text{sect mod}$$

(where sect mod = $\pi \times D^3/32$).

From which the minimum pole diameter D is = 292.9 mm.

Diameter at 1.5 m from butt (assuming taper 11 mm/m) = 296.2 mm. This is less than the specified minimum diameter of 305 mm for a stout pole within BS 1990, and hence a stout grade pole is okay.

At an angle or terminal positions the supports are stayed. In addition to the bending due to the wind loading, the pole must also act as a strut due to the vertical component of the stay tension (Figure 6.3).

In Figure 6.3, the conductors are represented by the forces T. These combine vectorially to give a resultant force, P. This has to be balanced by the force in the pole, V, and the stay wire, S.

6.2.2.3 Crippling load of struts (as applied to wood poles)

A strut refers to a member that is long in comparison with its cross-sectional dimensions. This describes a wood pole used as a direct support or as a stay. This will fail due to buckling before the compressive stresses reach yield point. The load that will cause buckling is given by Euler's theory. This can only be approximate in the case of wood poles. Wood poles have a varying diameter or taper, there are imperfections along their length (e.g. knots in the wood) and the actual strength of the wood may vary along the length. The formula used has taken this into account and has an inbuilt factor of safety.

Although a stayed pole does not strictly have both ends pinned, the formula for a beam pinned at both ends is used. However, difficulty arises with the effective length of the beam. Is the top end to be taken as the stay attachment point or the effective point of conductor attachment? The pole diameter is another contentious

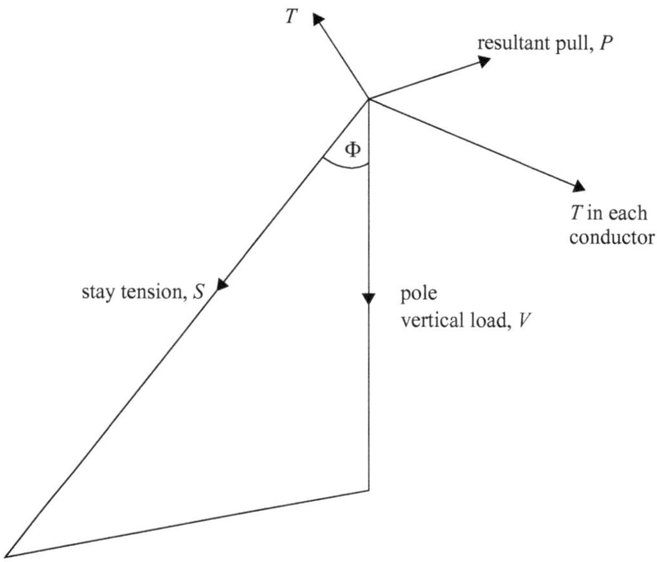

Figure 6.3 *Stay tension*

T = conductor tension in N
Φ = angle between stay and pole
S = tension in stay in N
V = vertical load on pole
P = horizontal pull at pole top and $= T$ for terminal load
then $S = P/\sin\Phi$ N
and $V = S\cos\Phi$ N

issue. Where should the diameter be taken? If an average is to be used, where and what should be averaged?

There can thus be large discrepancies depending on which parameters are used. The crippling load on a pole is calculated from Euler's theory where the length l is taken as the distance between the pole top and one and a half metres from the butt and the diameter as the average of the diameters at these points, since these are readily available.

Euler's theory

The formula commonly used to calculate the crippling load (P) for poles is:

$$P = \frac{\pi^2 EI}{l^2}\,\text{N} \tag{6.5}$$

where l is the length of strut in mm, E is the modulus of elasticity of pole (average value 10 054 N/mm^2) and I is the moment of inertia of a circular section ($\pi \times D^4/64$) mm^4.

6.3 Alternative wood pole support structures

There are many different types of pole. If the above calculations give loads that cannot be resisted by conventionally-sized stout poles, it may be useful to look at the alternatives. Some alternative designs are an H pole, A pole or Rutter pole, a design where two single poles are paralleled and joined together using a number of shaped wedges and long bolts. When a Rutter pole, a braced H pole or an A pole with a brace bar half way up the pole is used, then Euler's formula becomes:

$$P = \frac{4\pi^2 EI}{l^2} \tag{6.6}$$

There is an immediate fourfold increase in buckling strength and buckling need no longer be taken into account.

6.4 Cross-arms

6.4.1 General

The traditional method of supporting conductors was to use a cross-arm to hold the insulators that held the conductors. These cross-arms came in many shapes and sizes and indeed are being replaced by long rod insulators in some modern designs. Although we only deal with the traditional designs in this section, the principles of design also relate to modern insulators/cross-arms.

6.4.2 Design of cross-arms

Cross-arms for section angle and terminal supports are designed for horizontal loads imposed by the tension in the conductors. The normal maximum tension for 11 kV wood pole lines is of the order of 18 kN. It is not normal to design cross-arms for broken wire conditions since the movement of the pole under those conditions is generally sufficient to reduce the tensions to acceptable limits. This is not to say that under certain conditions the cross-arms cannot become twisted and occasionally the pole tops can be damaged. Modern designs, such as ENATS 43-40, fit failure containment devices in an attempt to reduce pole-top damage.

A cross-arm mounted on a single support can be treated as two cantilevers fixed at the support (Figure 6.4). For an intermediate cross-arm with pin insulators there are three bending moments that have to be considered:

1 Bending moment due to the weight W of the ice-loaded conductor acting vertically at a distance l from the pole centre.
2 Bending moment due to the tension from the wind-loaded conductor P acting at a distance X above the cross-arm.
3 Bending moment due to downthrust or uplift depending on the line profile.

In Figure 6.4, the force, P, is from the conductor which is suspended above the cross-arm by the insulator of height, x. This gives a downward force, W.

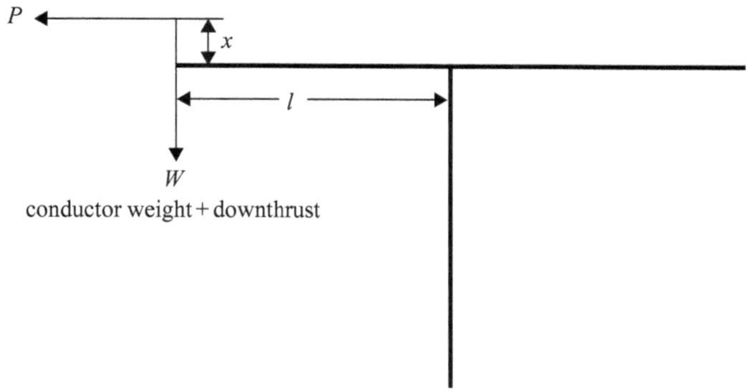

Figure 6.4 *Calculate maximum stress of conductor*

Let:

F = maximum allowable stress in the material

BM = bending moment

Z = section modulus

To calculate maximum working stress

$$F = BM/Z$$

Again, a worked example is useful to understand these calculations.

Example

To calculate the section modulus of the steel needed to support the line in Figure 6.4.

maximum allowable stress F = ultimate strength of material/factor of safety

maximum working stress $F = \frac{BM}{Z}$ therefore $Z = \frac{BM}{F}$

The cross-arm steelwork is taken as type S275JR to BS EN 10025 (i.e. BS 4360 grade 43B). In this case the maximum working stress is 43 000 N/cm^2.

If the factor of safety to be used is 2.5, then the maximum allowable working stress is

$$\frac{43\,000}{2.5} = 17\,200\ \text{N/cm}^2$$

If the maximum conductor wind loading P = 1500 N and acts at a distance x = 250 mm, and the maximum ice-loaded conductor weight W also = 1500 N, acting at a distance L = 1000 mm, then the maximum bending moment is calculated as:

$$1500 \times 25 + 1500 \times 100 = 187\,500\ \text{Ncm}$$

Minimum required section modulus is:

$$Z = \frac{187\,500}{17\,200} = 10.9\,\text{cm}^3$$

It is normal practice to allow, say, a ten per cent increase in the required section modulus to allow for the effect of mounting holes through the steel section, which effectively reduce the section modulus, and hence assume a minimum required section modulus of $12.0\,\text{cm}^3$. Steel milling companies publish tables of properties of their steel sections, and, in the case of cross-arms, these will generally be fabricated from angle millings. Typical tables indicate a range of steel angle sizes that would be suitable.

$80 \times 80 \times 10$ mm equal angle has a modulus of $12.6\,\text{cm}^3$, and $80 \times 60 \times 8$ mm unequal angle with the long side vertical has a modulus of $12.1\,\text{cm}^3$ but only has a modulus of 7.09 with the short side vertical. There are a number of different steel sections that could be employed, although the lighter sections are recommended for handling issues.

The same methods give the section (or elastic) modulus for channels and box section steel.

When designing a terminal cross-arm, the cross-arm must be able to withstand the stress due to the conductor tension as well as the weight of the ice-covered conductor. Since these forces act at 90° to each other, the cross-arm should be designed such that the actual longitudinal stress divided by the permissible longitudinal stress plus the actual vertical stress divided by the permissible vertical stress is less than or equal to 1.00.

6.4.3 Formulae for maximum bending moment

Figure 6.5 shows a standard intermediate cross-arm. The bending moments are shown in Table 6.1. Forces acting in different directions are indicated in the table as +ve or −ve; if in uplift it would be +ve. The formula for maximum bending moment = $Wx - 0.75Wy + 0.5Ph$ or, in some cases, $Wx - Wy + Ph$.

In Figure 6.5 the forces can be resolved vertically and horizontally at A and B:

$$A_{\text{horiz}} + B_{\text{horiz}} = 3P$$

$$A_{\text{vert}} + B_{\text{vert}} = 3W$$

Assume that the vertical forces at A and B are equal at 1.5 W. Moments about B for horizontal components only are:

$$A_{\text{horiz}} \times Z = 3P(h + Z)$$

$$A_{\text{horiz}} = 3P(h + Z)/Z$$

$$B_{\text{horiz}} = -3P(h)/Z$$

Compression in strut DB

$$= 0.5B_{\text{vert}}\sqrt{y^2 + Z^2}/Z - 0.5B_{\text{horiz}}\sqrt{y^2 + Z^2}/y$$

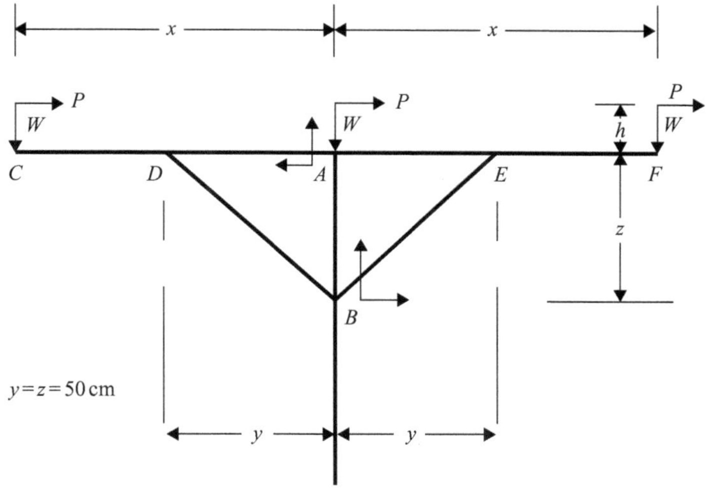

Figure 6.5 *Schematic intermediate cross-arm*

Table 6.1 *Bending moments*

C^+	$-Ph$
D	$W(x-y)-Ph$
A^-	$Wx-0.75Wy+0.5Ph$
A^+	$Wx-0.75Wy-0.5Ph$
E	$Wx-Wy+Ph$
F^-	Ph
F^+	0

Substituting for B, compression in strut DB

$$= 0.75W\sqrt{y^2+Z^2}/Z - 1.5P(h)\sqrt{y^2+Z^2}/y$$

The vertical component

$$= 0.75W - \frac{1.5Ph}{y}$$

and the horizontal component

$$= \frac{0.75Wy}{Z} - \frac{1.5Ph}{Z}$$

In strut EB the vertical component

$$= 0.75W + \frac{1.5Ph}{y}$$

and the horizontal component

$$= \frac{0.75Wy}{Z} + \frac{1.5Ph}{Z}$$

Compression in strut EB

$$= \frac{0.75W\sqrt{y^2 + Z^2}}{Z} + \frac{1.5Ph\sqrt{y^2 + Z^2}}{Zy}$$

When designing for large conductors there may be a risk of considerable deflection of the cross-arm. If this is the case, an additional stay may have to be fitted to the ends of the cross-arm. This imposes a considerable vertical loading on the cross-arm which, together with the weight of the ice-loaded conductors, may become the critical factor. In this instance, the vertical loading will give the minimum cross-arm size.

6.5 Stays and stay loading

The design of any staying arrangement must take into account the following items.

6.5.1 Maximum working tension of stay wire

Where a traditional stay arrangement is used, then the weakest item in the stay assembly is the stay wire, assuming that the soil is of adequate property to resist the load in the staywire. If 7/4.00 mm grade 700 wire is used then the ultimate tensile strength (UTS) is 61.6 kN. The factor of safety normally used with stays is 2.5. This gives a normal maximum working load of 24.64 kN.

6.5.2 Stay foundation

The vertical component of stay tension (stay wire tension × cos stay angle) must be equal to or less than the frustum weight of soil that is resisting the stay block pulling out of the ground. For a stay block 850 × 250 × 125 mm, the volume of the soil frustum is 4 m^3, assuming a 2500 mm stay rod in normal soil (frusta angle 30°) and a stay angle of 45°. The density of soil is generally taken as 1595 kg/m^3 (as noted in ENATS 43-91), which equates to a sandy soil. Hence the resistance of the soil would be 62.5 kN compared with a vertical component of stay tension of $61.6 \times \cos 45° = 43.6$ kN, hence okay.

6.5.3 Pole crippling capability

For a 10 m medium pole with minimum top diameter the crippling capability is given as 76.39 kN in BS 1990. Using a factor of safety of 2.5 this gives us a maximum working load of 30.56 kN. It should be noted that as the stay angle decreases (perhaps due to some unavoidable obstacle), the load in both the stay and pole increases. The resultant loads in the pole will increase by a factor 1/tan (stay angle) to which conductor downpull, cross-arm steelwork and insulator weight need to be added.

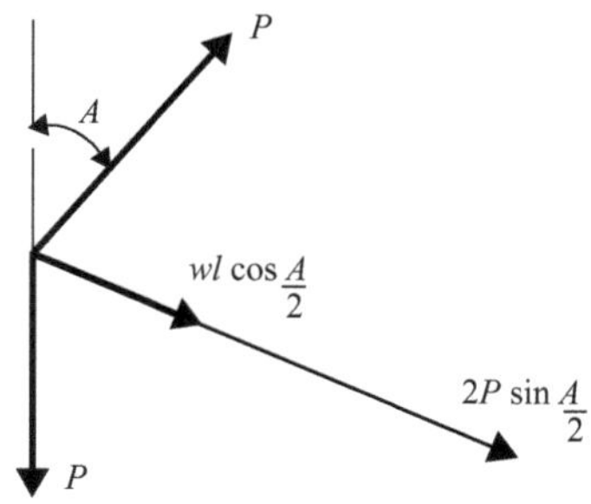

Figure 6.6 Resultant forces at an angle (section) pole

6.5.4 Resultant pull on the pole

The resultant pull on the pole includes windage on the area of pole above ground and the windage on any pole top equipment.

Where modern alternative stays, such as the duckbill stay anchor (so named because the end, which is pointed to allow easy penetration before rotating back in the soil to bear against undisturbed soil, is shaped very much like a duck's bill), are used then reduced factors of safety may be applied in certain situations since the actual holding capability can be tested.

For an angle pole (Figure 6.6):

A = line angle of deviation

P = total max working tension of all conductors

l = weight span = half the sum of adjacent spans

w = wind load for all conductor per metre

Pull on pole top $= 2P\sin A/2 + wl\cos A/2 +$ windage on pole + windage on pole-top equipment. The windage is small by comparison and $\cos A/2$ can be made equal to 1. Pull on pole top $= 2P\sin A/2 + wl$.

For a terminal pole (Figure 6.7):

P = total conductor tension of all conductors

tension in stay = pull on pole/$\sin\phi$

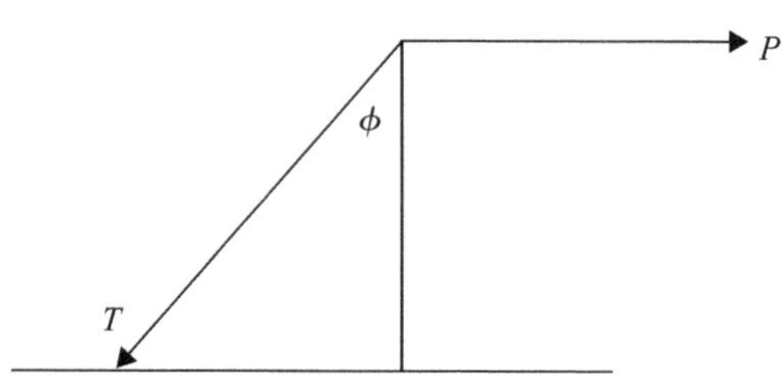

Figure 6.7 Terminal pole

6.6 Summary

This chapter was intended to give an introduction to some aspects of wood pole design. It does not pretend to be a complete guide, and in view of the computer programs available today it is in many ways an anachronism. However, it does show how all aspects of pole design for both one-off and standard designs can be produced from first principles. A basic understanding of the engineering calculations enables the output from computer programs to be checked.

6.7 Further reading

McCOMBE, J. and HAIGH, F. R.: 'Overhead line practice' (MacDonald, London, 1966)

GOVEN, T.: 'Electric power transmission systems engineering – analysis and design' (Electricity Association, London, 1988)

MORECOMBE, W.: 'Overhead power lines' (Chapman and Hall)

SMITH, S.: 'Study of overhead distribution lines and their design parameters' (Energy Networks Association, London)

Chapter 7
Weather loads, conductor sags and tensions

7.1 Scope

The next two chapters look at conductors – the workhorse of the overhead line. Chapter 8 looks at the electrical choice for conductors and how to calculate what is required. This chapter, however, continues with the mechanical theme of chapter 6. The conductor that goes up must maintain its statutory clearance to ground for the next 40 or 50 years, whether cold or up to its maximum allowable operating temperature (normally 50 °C but now often higher). Mechanical and metallurgical creep cause a conductor to stretch and so slacken off between the poles and this must be allowed for.

Conductors also take wind and ice loads and, within reason, should not break or be strained beyond their elastic limit under severe weather conditions. There is, therefore, a need to make further allowances e.g. use factors of safety.

Finally, not all lines are on level ground. Hills are always present and sag/tension calculations must take account of the fact that an intermediate pole may be on top of a hill or deep in a gully. This chapter therefore seeks to cover the most general aspects of conductor sag and tension calculations, including the weather loads and uneven ground and compares manual methods of line design with computer software design packages.

7.2 Conductor loadings

7.2.1 General

In order to accurately calculate structure loadings and line clearances, sag/tension calculations will need to be undertaken which reflect the weather conditions for the line route in question. A large number of lines in the UK have been designed to the old Overhead Line Regulations 1970 (or its precursors) that included a factored design approach, namely, 760 N/m^2 (16 lbs/ft^2) wind on bare conductor with a factor of safety of 2.5 on the conductor breaking load, for conductors up to 50 mm^2 equivalent aluminium area, and 380 N/m^2 (8 lbs/ft^2) wind on ice-loaded conductor with a factor of safety of 2.0 for larger conductors. Ice loading was specified as a minimum of 9.5 mm in the OHL regulations (3/8″), although steel tower lines have historically employed a larger ice accretion of 12.5 mm (1/2″), all assumed acting at −5.6 °C.

These national loadings have not proven satisfactory for certain areas of the UK, resulting in a higher degree of line failure than anticipated. A more sensible approach is to design lines for the particular climate along the line route in question. In this respect, the principles included in BS 8100 have been incorporated into the NNAs of the newly published BS EN 50341 as the UK consideration of a probabilistic design (general approach) for steel tower lines operating at voltages greater than 45 kV.

For lines below 45 kV, it has now been agreed that 380 N/m^2 wind and 9.5 mm radial ice and 570 N/m^2 wind with 12.5 mm radial ice be employed for a normal deterministic design (empirical approach), subject to the altitude of the line. Further details are included in BS EN 50423-3-9 (lines below 45 kV) and/or ENATS 43-40 Issue 2.

For the purposes of this chapter, however, the simpler OHL Regulations 1970 loading regime has been assumed, although the principles that follow equally apply where any of the weather loadings from the above design codes are employed.

Once the basic ice accretion thickness is known, the unit weight of an ice-loaded conductor can be calculated, assuming, in this case, an ice density of 0.913 g/cm^3 (i.e. non-aerated, glazed ice):

$$\text{ice loaded weight} = W_B + 0.717/1000 \times 2R \times (2R + 2d) \text{ kg/m} \qquad (7.1)$$

where W_B is bare conductor weight (kg/m), R is radial thickness of ice (mm) and d is bare diameter of conductor (mm) (for the derivation of this formula, see section 7.12, Appendix A).

Maximum wind load on the ice-loaded conductor is:

$$P(d + 2R)/1000 \text{ (kg/m)} \qquad (7.2)$$

where P is the gust wind pressure (i.e. 77.5 or 38.75 kg/m^2 as necessary).

7.2.2 *Combined loadings*

The gust wind speed is a horizontal load, and the ice-loaded conductor is a vertical load hence, in order to rationalise these loads for sag/tension calculations, we must now calculate the vector sum of loads that the conductor will experience in the form of a maximum conductor resultant:

$$\text{MCR} = \sqrt{}([\text{ice-loaded weight}]^2 + [\text{maximum wind load}]^2) \text{ (kg/m)}$$

The MCR is the augmented load of the conductor due to wind and ice that is used in the conductor change of state calculation as a limiting criteria (i.e. in this case, the maximum conductor tension that is permitted for a line based on the applied wind and ice loadings).

7.3 Tension limits

A conductor is strung taking into account the maximum tension it will see during its life. This is generally known as the maximum working tension (MWT), and assumes that maximum wind and ice accretion occur at the same time, normally at minimum temperature.

A second tension limit is often employed to avoid harmful Aeolian vibration occurring during normal ambient temperatures and is generally related to a percentage of the ultimate strength of the conductor (section 7.4).

In addition, for steel towers lines it is common practice to include an erection limit within the sag/tension calculation. The erection limiting tension is that tension to which the conductor will be erected and includes any overtension/temperature shift to compensate for creep (section 7.8). It is generally assumed that a conductor would not be erected at a temperature lower than 0 °C (in still air) and hence the tension limit would be expressed at, say, −20 °C for a 20 °C temperature shift or, say, 1.1 × design tension at 0 °C for ten per cent overtension.

Depending on the span lengths (or equivalent span lengths) selected, it may be that one, two or all three limiting tensions will rule depending on the equivalent span selected (i.e. erection limit = 0–60 m, EDT = 60–150 m, MWT > 150 m).

7.4 Vibration limit

Aeolian vibration is a wind-induced oscillation. The vibration is caused by low-wind-speed eddies on the leeward side of the conductor that swing from the upper to the lower side at regular intervals, the rate depending on the conductor diameter and the wind velocity. Vibration occurs if the frequency of the swing caused by the eddies matches with that of the resonant conductor frequency. A steady low-velocity crosswind is all that is required to sustain the eddies and generate a high-frequency (5–100 Hz) low-amplitude oscillation. The constant flexing of the conductor can lead to fretting of the strands and ultimate failure of the whole conductor. Since Aeolian vibration is related to conductor tension, it has been found that bare, stranded conductors will avoid damage if strung at a tension at or below certain limits in still air.

The conductor vulnerability to damage increases with conductor tension and this damage generally occurs at suspension or damper clamps where the conductor is forced into a node. This can result in many bending cycles of high enough amplitude to initiate fractures in the conductor strands. Once these develop sufficiently (normally about 1/3 diameter) the strands tend to break under elastic tensile stress. To avoid this scenario, the historical technique has been to limit the conductor tension to below a specific value related to the conductor material. These recommendations came initially from Zetterholm in a 1960 Cigré session paper and then later from Cigré SC22 WG04 in 1979. Even then, however, the effect of terrain was noted and allowance was made for hilly, as opposed to flat, terrain. In 1990, there was a move to have a fresh look at the situation as line failures were still occurring. In most parts of the world, vibration fatigue was the dominating life-limiting effect. In the UK, however, corrosion is a more important factor for aluminium-based conductors in coastal areas. This is covered in detail in chapter 8.

Cigré has set safe design tension limits to avoid the effects of Aeolian vibration and is currently investigating their validity. Historically, there are set tension limits for overhead line conductors to avoid failure due to Aeolian vibration fatigue. These limits have been based on a pure percentage of the conductor UTS value (Table 7.2). Historically, in the UK, however, slightly different values have been used such as 20 per cent UTS for aluminium, 33 per cent UTS for copper, 18 per cent for ACSR etc. (Table 7.2). However, there have been many line failures world-wide at tensions below recommended percentage UTS limits. Tables 7.1 and 7.2 give Cigré recommended values which are not necessarily followed in individual countries.

The everyday design stress (at 5 °C), or EDS, based on the percentage UTS concept, is the definition of a maximum tensile load that a conductor must be limited to at the temperature that the conductor is subject to for the longest period of time. Different values are given for single, damped and twin conductors. Data on the service life of conductors against EDS values (Cigré Task Force 04 of Cigré Standing Committee B2 (Overhead Lines) Working Group 11 (SCB2 WG11 data)) showed that these values were not a safe level, but only an indication of the rate of fatigue. Table 7.1 summarises world-wide data for 175 mm^2 Lynx ACSR conductor, for which the EDS value was 18 per cent UTS.

Table 7.1 Summary of Lynx performance and EDS

Service life (years)	Percentage of lines damaged	
	EDS < 18%	EDS > 18%
= 5	5.26	25.00
5 to 10	20.93	35.29
10 to 20	45.00	78.00
>20	58.93	91.67

Table 7.2 EDS recommendations (1960) in % UTS

Conductor type	No dampers	Dampers
Copper	26	
ACSR	18	24
Aluminium	17	
Aluminium alloy	18	26
Steel	11–13	

It is significant that 45 per cent of lines erected at <18 per cent UTS failed after service lives of less than 20 years.

The overall EDS recommendations are given in Table 7.2.

At distribution voltage levels dampers are rarely used, so the values in the first column of Table 7.2 were used. Over time it became obvious that the copper value was too low and the common value used is 33 per cent UTS.

Over the last few years, a new approach based on tension, conductor weight, terrain factors etc. has been followed by SCB2 WG11 TF04.

The first part of the task force research covered both an accumulation of field data and the modelling of Aeolian vibrations of single, undamped conductors, in order to define the vibration level and to assess the uncertainties connected to the present technology based on the energy balance principle. It was found that, even with tensions within the old safe limits, if the ratio between tension and cable unit weight (H/W) exceeds certain limits, Aeolian vibrations may still cause serious conductor and fitting damage. This was confirmed by applying the new vibration factors to known failure scenarios collected from world-wide data.

7.4.1 Effect of terrain

Calculations on wind power input assume a steady wind along the whole span. Gustiness or variations in the wind speed along the span can significantly reduce the wind power input. This implies that local ground conditions can have a significant effect on the lowest phase conductors on tower lines, reducing power input and allowing higher tension limits. The turbulence is defined by the root mean square variation of the wind speed about the mean wind speed and the turbulence intensity as the ratio of this to the mean wind speed, expressed as a percentage. Turbulence is therefore strongly dependent on local ground cover – trees, bushes, buildings etc. Table 7.3 gives a rough classification of ground roughness, although Cigré gives a more refined classification [1].

Table 7.3 Turbulence intensity against terrain roughness

Terrain factor	Terrain description	Turbulence (%)
1	open water or desert, snow cover, no trees, no obstructions	8
2	open, flat rural areas with no obstructions and few and low obstacles	15
3	open, flat or undulating; low-density housing, open woodland with hedgerows and small trees, prairie, tundra	22
4	built-up areas with some trees and buildings, e.g. suburbs, small towns, woodlands and shrubs, broken country with large trees, small fields with hedges	30

The values in Table 7.3 can be reduced in low wind speed (<10 m/s) and cold conditions, as the atmosphere tends to stratify closer to the ground. At higher wind speeds, the values are more consistent and reliable.

One of the main differences between the original EDS concept and the current thinking on safe design tension is the effect of ground roughness on wind turbulence and hence on the wind power input to the conductor. Fluctuations of wind speed with time and also with position along the span can significantly reduce power input, as the resonant frequency cannot respond quickly to wind speed variations. The concept of turbulence intensity has been explained. The turbulence depends on local effects (trees, buildings etc.) and ground cover. Larger features, such as hills and ridges, shape the flow on a large scale and, although wind speeds may change (e.g. due to funnelling), the actual turbulence intensity may reduce.

Significant turbulence (and higher effective terrain factors) can be generated by trees or buildings of 5–10 m height on the prevailing wind side of a span.

The current recommendations from Task Force 04 [1] are:

1 The maximum safe design tension for undamped single conductors is given in Table 7.4 in terms of H/W for four terrain categories. The factor, H (design tension), is evaluated for the mean temperature of the coolest month (MTCM) and is the initial design tension, not the actual tension *in situ*.
2 The tension values (see chapter 8 for conductor abbreviations) apply to:
   - A1 (AAC)
   - A2, A3 (AAAC)
   - A1/A2 or A1/A3 (ACAR)
   - A1/St (ACSR).
3 If there is any doubt as to the terrain category the recommendation is to choose the next lowest H/W value.
4 Spans exposed to corrosion or pollutants may have reduced self-damping and so lower fatigue endurance.
5 The use of cushioned clamps or helical fittings can justify the use of higher H/W values than in Table 7.4 (no specific value is recommended, however).
6 If lines are strung at higher than recommended values of H/W, then field vibration measurements are recommended.

Table 7.4 Recommended H/W values for terrain factors

Terrain factor	H/W
1	1000
2	1125
3	1225
4	1425

The values in Table 7.4 can be converted to percentage UTS to give allowable tensions for a range of conductors at different span lengths. The values are mostly well below those normally used in the UK.

The data are based on winds and terrain effects at 10 m above ground level. Most medium voltage lines are liable to be so close to the ground and over fields and hedgerows that terrain factors 2 and 3 are most likely applicable. Conductors strung at the original Cigré recommended tensions could be assumed to have a lifetime of 50 years for aluminium based and 60 years for copper based.

Vibration dampers and spacer/dampers can be used to dissipate harmful vibrations on long-span transmission lines. It is not possible to keep a conductor at precisely these tensions due to temperature variations. However, during spring and autumn, some mean temperature (taken as 5 °C in the UK) will be evident and this is usually adopted as an everyday temperature for these vibration-limiting everyday tensions (EDT). During summer, the tensions will be lower due to the higher seasonal temperatures.

Unfortunately, compacted conductors, particularly covered conductors, are more prone to vibration due to the lack of strand interstices that would otherwise introduce self-damping and reduce the effect of the wind. Experience has shown that small size, covered, aluminium-based conductors need a vibration limit of 10–12 per cent of the calculated conductor breaking load to avoid strand damage, unless vibration dampers or other spiral devices are employed. Such a low conductor tension gives a proportionally large sag, particularly at the maximum operating temperature.

Where higher everyday tensions are needed to minimise conductor sag, spiral impact (Figure 7.1) or stockbridge-type (shaped like small dumb-bells attached

Figure 7.1 Spiral vibration dampers (SVD) on a CC

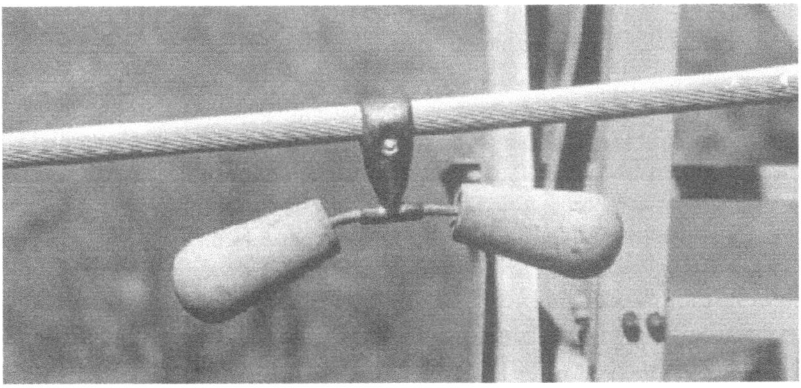

Figure 7.2 Stockbridge vibration damper

to the line) (Figure 7.2) vibration dampers should be used. These effectively take the energy out of the oscillations.

7.5 Sag/tension calculations

The principles of sag/tension calculations were explained in section 4.2.6 but now the detailed calculations are provided.

7.5.1 Catenary method

The most accurate calculation for sags and tensions is the catenary method that describes the shape of the sag curve produced.

Rigid supports at points A and B, as shown in Figure 7.3, support a conductor ACB (on level ground).

S = true conductor length

L = span length

D = sag

T = horizontal tension at C

W = weight per unit length

c = distance between the origin and point C

The equation of the catenary from C to B is:

$$y = c \cosh x/c$$

The origin is chosen so that, by definition:

$$c = T/W$$

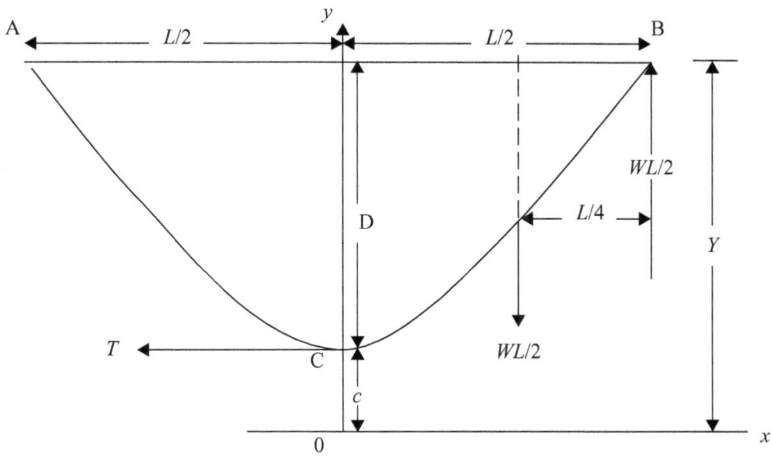

Figure 7.3 Level span

The true conductor length from C to B is given by $S/2 = c \sinh L/(2c)$ and hence the true conductor length is:

$$S = 2T/W \sinh[WL/(2T)] \tag{7.3}$$

The conductor sag $D = Y - c$ is:

$$D = T/W\{\cosh[WL/(2T)] - 1\} \tag{7.4}$$

Both of the above equations may be expanded, giving:

$$\begin{aligned} \text{true length, } S = {} & 2T/W\{[WL/(2T)] + [WL/(2T)]^3 1/3! \\ & + [WL/(2T)]^5 1/5! + \cdots\} \\ & \times L + W^2L^3/(24T^2) + W^4L^5/(1920T^4) + \cdots \end{aligned} \tag{7.5}$$

and

$$\begin{aligned} \text{sag } D = {} & T/W\{[WL/(2T)]^2 1/2! + [WL/(2T)]^4 1/4! \\ & + [WL/(2T)]^6 1/6! + \cdots\} WL^2/8T + W/(6T)(WL^2/8T)^2 \\ & + 0.4 \times [W/(6T)]^2(WL^2/8T)^3 + \cdots \end{aligned} \tag{7.6}$$

It is important to note that T in the above formulae will indicate the maximum horizontal tension at the point in the catenary where the conductor slope is horizontal. At the ends of the level span, the total conductor tension is equal to the horizontal component plus the conductor weight per unit length multiplied by the sag, as shown in the following:

$$\text{total}_T = \sqrt{[(MWT_H)^2 + (MCR \times L/2)^2]} \tag{7.7}$$

For very long spans, care must be taken to distinguish between horizontal and total tension. All following references to tension refer to horizontal tension, unless otherwise stated.

7.5.2 Parabolic method

An easier and simpler technique for calculating sag/tensions is to assume that the conductor forms a parabola and not a catenary. A parabola closely resembles a catenary. As long as the span is less than 300 m and has a sag less than five per cent of the span length, the error between the catenary and parabolic methods approximates to 0.5 per cent, which, for wood pole lines, is not significant. If the sag curve is taken as a parabola, therefore, then taking moments about B in Figure 7.3, the tension multiplied by the sag will be equal to half the weight of the conductor. This is assumed to be acting at a point midway between C and B, i.e.:

$$T \times D = WL/2 \times L/4$$

and hence:

$$\text{sag, } D = WL^2/(8T) \tag{7.8}$$

It can be shown that true conductor length equals:

$$S = L + W^2L^3/(24T^2) \tag{7.9}$$

and substituting for sag, D:

$$S = L + 8D^2/(3L) \tag{7.10}$$

Note that the parabolic equations (7.8) and (7.9) are the same as the first terms of the expansions of the catenary equations (7.6) and (7.5), respectively.

The difference between span length and true conductor length is often called slack. Slack has units of length, but is often expressed as a percentage relative to the span length. For a given tension/weight ratio, slack is related to the cube of span length and to the square of the sag for a given span.

We now have two loading criteria:

1 wind and ice for the maximum working tension at −5.6 °C
2 still air for the everyday tension, normally at 5 °C.

A tension calculation must now be undertaken to determine which of these criteria will rule (i.e. one of these criteria will be at its maximum, and so dominate, and the other will be lower than its maximum). The following change of state equation, derived from parabolic equations (7.8) and (7.9), is usually sufficiently accurate:

$$\text{initial conductor length} \pm \text{change in length} = \text{final conductor length}$$

Hence:

$$L + W_1^2L^3/(24T_1^2) \pm [\alpha L(t_2 - t_1) + (T_2 - T_1)/(aE) \times L] = L + W_2^2L^3/(24T_2^2) \tag{7.11}$$

Subtracting L from each side and multiplying throughout by aE/L:

$$\pm Ea\alpha(t_2 - t_1) + W_1^2L^2Ea/(24T_1^2) - T_1 = W_2^2L^2Ea/(24T_2^2) - T_2$$

This is normally re-arranged as follows:

$$T_2^2[T_2 + W_1^2L^2Ea/(24T_1^2) - T_1 \pm Ea\alpha(t_2 - t_1)] = W_2^2L^2Ea/24 \qquad (7.12)$$

where:

E = modulus of elasticity of conductor (kg/mm^2)

a = total cross sectional area (mm^2)

α = coefficient of linear expansion (m extension/m of span/°C)

t_1 = initial temperature (°C)

t_2 = final temperature (°C)

W_1 = initial unit weight (if wind and ice loaded, this would be MCR) (kg/m)

W_2 = final unit weight (if in still air, this would be the bare weight) (kg/m)

L = span length (m)

T_1 = initial tension (kg) at t_1

T_2 = final tension (kg) at t_2

The conductor constants (i.e. a, E, α and W_2) can be found in any conductor manufacturer's catalogue.

The sag is as equation (7.8).

Equation (7.12) reduces to the form below and can be solved for T_2 by trial and error using a pocket calculator, although this can be somewhat tedious (but see section 7.5.3):

$$T_2^3 \pm KT_2^2 = N$$

A similar change of state formula is available for catenary calculations:

$$T_2/W_2[2\sinh LW_2/(2T_2)] - L[T_2/(Ea) + \alpha t_2]$$
$$= T_1/W_1[2\sinh LW_1/(2T_1)] - L[T_1/(Ea) + \alpha t_1]$$

and sag:

$$D = T_2/W_2[\cosh LW_2/(2T_2) - 1]$$

The tension equation reduces very little to:

$$T_2/A[2\sinh(B/T_2)] - CT_2 = N \qquad (7.13)$$

The goal-seek routine in spreadsheet packages such as Excel is of help, but this is not very automated. Also, for line sections with suspension insulators, this equation is only true for still-air conditions where EDT rules. Where values of W_1 are with wind and ice, separate calculations need to be undertaken to resolve the vertical and transverse component forces in the deflected plane of the conductor before applying the catenary equations above.

Once it has been determined which loading criteria rule, a second calculation needs to be undertaken using the ruling criteria to determine the tension for maximum sag in still air, which usually occurs at maximum conductor temperature, and hence determine the minimum ground clearance at that temperature. In certain climates, the maximum sag may not occur at the maximum conductor design temperature. It may be that an ice-loaded conductor at $-5.6\,^{\circ}\mathrm{C}$ in still air will sag more than a bare conductor at 50 °C in still air. It is therefore important to undertake a further sag/tension calculation to investigate this.

Conductors in this country have, historically, operated with a maximum temperature of 50 °C due mainly to the thermal constraints of the line fittings employed at that time. With the introduction of new materials and techniques it is possible to operate overhead line conductors up to 90 °C, beyond which annealing of aluminium-based conductors begins to take place. A usual maximum now employed, particularly overseas, is 75 °C and is chosen due to the higher ambient temperatures in those countries. Clearly, the smaller differential between everyday and maximum temperatures will restrict the continuous operating current that can be transmitted. To overcome the normal annealing constraints, NGT has commenced using gap conductors (**G**apped **Z**-shaped strand high **T**emperature **A**luminium **C**onductor **S**teel **R**einforced (GZTACSR)) where superheat-resistant aluminium alloy is supported by an extra high strength galvanised steel wire core. In the NGT case, the maximum operating temperature of 620 mm^2 Matthew GZTACSR (same diameter as 500 mm^2 Rubus) can be as high as 160 °C for the same sag as standard 400 mm^2 Zebra ACSR operating at 50 °C and with marginal increase in line tension (see Cigré 22-202).

7.5.3 *Newton–Raphson iteration*

Clearly, performing a manual iteration to the basic parabolic cubic equation would be very tedious, particularly if a number of limiting conditions, spans or temperatures were required.

Fortunately, there is mathematical help at hand for calculating any power law equation, including the parabolic equation, using the Newton–Raphson iteration law. If the expression is re-arranged to sum to zero, and divided by its differential, a value, V, will be obtained, if an initially assumed value of T_2 is used throughout. This value, V, is then used to modify the initially assumed value of T_2 and the calculation repeated to iterate for T_2:

$$\pm V = (T_2^3 \pm K T_2^2 - N)/(3T_2^2 \pm 2K T_2) \tag{7.14}$$

Once V tends towards zero, the value of T_2 will be the actual root of the equation. From experience, eight or nine iterations are required until zero is reached for cubic tension calculations and, hence, only 16 or 18 spreadsheet cells need be used to accurately, and automatically, solve for T_2. The following example for 100 mm^2 AAAC Oak at 50 °C indicates the process:

$$T^3 + 123.085T^2 = 50\,100\,972.3$$

Table 7.5 Newton–Raphson iteration

Value of T	1788.1	1788.1 − 604.12 = 1183.98	1183.98 − 396.31 = 787.67	787.67 − 250.57 = 537.10	537.10 − 140.68 = 396.42	396.42 − 55.43 = 340.99	340.99 − 8.92 = 332.07	332.07 − 0.22 = 331.85
Iteration	604.12	396.3	250.57	140.68	55.43	8.92	0.22	0.00013

hence:

$$T^3 + 123.085T^2 - 50\,100\,972.3 = 0$$

Newton–Raphson formula is:

$$\frac{(f)}{dy/dx(f)} = 0$$

For the cubic tension formula above:

$$\frac{T^3 + 123.085T^2 - 50\,100\,972.3}{3T^2 + (2 \times 123.085 \times T)}$$

If we assign an initial tension value of, say, UTS/2.0, i.e. 1788.1 kgf, then we get the figures given in Table 7.5. From Table 7.5, the correct root of the equation, 331.85 kgf, is quickly ascertained.

A spreadsheet for sag and tension calculations is also of immense assistance where it is necessary to interactively amend data to obtain a specific sag or tension result. For example, where an existing conductor is being re-strung with another, it is obviously necessary to ensure that the maximum sag value is not exceeded. The EDT value may be revised until the required basic sag value is achieved, the spreadsheet providing the amended MWT value for assessing strength of components.

7.6 Conductor slack

As stated earlier, conductor slack is the difference between the true or actual length of the suspended conductor and the horizontal span length. By varying the value of t_2 in equation (7.11), a table of tensions can be developed indicating the increase/decrease due to temperature/stress changes as the conductor temperature changes. Table 7.6 has been developed for the same 100 mm^2 AAAC Oak conductor considered previously for a horizontal span of 125 m.

The interesting point to observe here is the very minor change in conductor length that is necessary to increase the sag of the conductor by, say, 1 m (i.e. 2.304 − 1.298 = 1.006 m, in this case between 25 °C and 65 °C). The corresponding increase in true conductor length is only 125.1132 − 125.0359 = 77 mm (3 in). It is therefore imperative that, whenever midspan joints or new tension joints are introduced into

Table 7.6 Actual conductor length for 100 mm² AAAC Oak conductor

Still air tension (kgf)	Temp. (°C)	True cond. length at 5 °C	Change in length (m)			True cond. length at temp.	125 m span sag (m)
			Due to stress	Due to temp.	Total change		
850.96	−5	125.01712	0.0237	−0.0288	−0.0050	125.0121	0.753
781.54	0	125.01712	0.0116	−0.0144	−0.0028	125.0143	0.820
715.20	5	125.0171	0.0000	0.0000	0.0000	125.0171	0.896
594.49	15	125.01712	−0.0211	0.0288	0.0077	125.0248	1.078
493.74	25	125.01712	−0.0387	0.0575	0.0188	125.0359	1.298
414.83	35	125.01712	−0.0525	0.0863	0.0338	125.0509	1.544
355.53	45	125.01712	−0.0629	0.1150	0.0521	125.0693	1.802
331.85	50	125.01712	−0.0670	0.1294	0.0624	125.0795	1.930
311.38	55	125.01712	−0.0706	0.1438	0.0732	125.0903	2.057
278.07	65	125.01712	−0.0764	0.1725	0.0961	125.1132	2.304
252.35	75	125.01712	−0.0809	0.2013	0.1204	125.1375	2.539

a conductor, the conductor measurements are undertaken accurately to avoid any under/over-sag.

7.7 Span length

7.7.1 Level spans

The above models assume a single span, L, of some fixed length. In practice, of course, a line is made up of many spans between section structures (i.e. where a conductor is rigidly terminated). There may be a section of, say, 20 spans with varying actual span lengths. For wood pole lines, it is normal to identify one basic span for a conductor, perhaps 110 m, which can then be used for sag/tension calculations. Providing all spans approximate to this value, there is usually no problem. However, if span lengths in a section vary wildly, a more accurate assessment of a weighted span length needs to be calculated. This is known as an equivalent span and is calculated as follows:

$$\text{equivalent span} = \sqrt{[(L_1^3 + L_2^3 + L_3^3 + L_4^3 + L_5^3 \cdots)/\text{section length}]} \quad (7.15)$$

The derivation for the above equation can be found in prIEC 61989 [2].

The equivalent or ruling-span method for determining sags and tensions is used on higher voltage wood pole lines (including trident) and all steel tower lines. It is common practice when designing such lines that three or more sag templates be employed to cover the expected range of equivalent spans encountered.

For 11 and 33 kV wood pole lines, it is normal to select an equivalent span that closely follows the majority of spans on the line in question and use this throughout the line.

Table 7.7 Sags and tensions for equivalent spans

Actual span	80 m equivalent span		110 m equivalent span		140 m equivalent span	
	Tension (kg)	Sag (m)	Tension (kg)	Sag (m)	Tension (kg)	Sag (m)
80 m at 50 °C	253.7	1.03	308.0	0.85	353.9	0.74
110 m at 50 °C	253.7	1.96	308.0	1.61	353.9	1.40
140 m at 50 °C	253.7	3.17	308.0	2.61	353.9	2.27
MWT at −5.6 °C	1193.2		1333.8		1464.9	

Although this provides a good deal of simplicity for erection staff, an over-sag problem will exist for short equivalent span sections.

Take, for example, a 100 mm^2 all aluminium alloy conductor (to BS EN 50182, old code name Oak) with the following equivalent spans – 80, 110 and 140 m – where the EDT rules in each case.

It can be seen from Table 7.7 that increasing the equivalent span leads to a smaller sag for the same actual span length. It might reasonably be assumed that as the equivalent span increases, the sag would also increase since $D \propto L^2$ from equation (7.8). This is certainly true for the actual spans (80, 110 and 140 m above), but for differing equivalent spans, as we have seen in equation (7.12), the extension of a conductor is related to variations in conductor temperature and variations in stress. For the same limiting tension (in this case EDT), the shorter equivalent spans will stretch proportionally more and consequently will result in more sag. For this reason, it is wise to make some small, additional allowance to the sag when considering ground clearance for wood pole lines employing a single equivalent span, where shorter equivalent spans are evident compared to the basic (or ruling) span.

In the examples given above, there will also be out of balance longitudinal loads at the section structures between differing OHL equivalent span lengths. This is because in all three equivalent spans, the same EDT value rules and as the temperature varies from the erection temperature, so will the relevant tension in each section.

Hence, for a 110 m basic span and an actual equivalent span (i.e. single-span section) of 80 m, there will be an out-of-balance load under MWT conditions of $(1333.8 - 1193.2) = 140.6$ kgf per conductor acting as an unstayed cantilever load towards the 110 m equivalent span section.

The breaking load for a typically sized 9.5 m stout pole to BS 1990 is approximately 1600 kgf, hence over a quarter of the pole strength has been eroded, during worst weather loadings, purely on this effect. This is the reason why many old lines were built with stays on section poles.

The increasing demand by landowners to prohibit stays on their land has led to many such stays being removed, on the basis that, if the line fails during an exceptional storm, the line can always be re-erected. It is the author's contention that reduced staffing may cause significant delays in line repair time, particularly where failed section structures are evident. If stays have to be avoided at these positions, larger diameter poles should be specified to take account of this loading.

7.7.2 Non-level spans

So far, we have only looked at level spans. It is necessary now to consider how sag is affected on non-level, i.e. hilly spans.

For calculation of a catenary over hilly ground, it can be shown from Figure 7.4 that CA and CB are each catenary curves of half spans X and $(L - X)$.

$$D_1 = T_H/w[\cosh(wX/T_H) - 1] \tag{7.16}$$

$$D_2 = T_H/w[\cosh(w[L - X]/T_H) - 1] \tag{7.17}$$

and

$$\text{total } T \text{ (at B)} = T_H \cosh[w(L - X)/T_H] \tag{7.18}$$

It should be noted that for tower lines, the calculation of a catenary over hilly ground is complicated by the displacement of the suspension insulator set and the reader is directed to the paper by Bradbury, Kuska and Tarr [3] for a full treatise of the required mathematics which addresses the displacement effect.

The equation of a parabola is of the type $Y = KZ^2$.

If D_3 = sag, and $L/2$ = half span length, then:

$$D_3 = K(L/2)^2 = KL^2/4 \quad \text{or} \quad K = 4D_3/L^2 \tag{7.19}$$

Inserting values for K:

$$Y = 4D_3/L^2 \times Z^2 \tag{7.20}$$

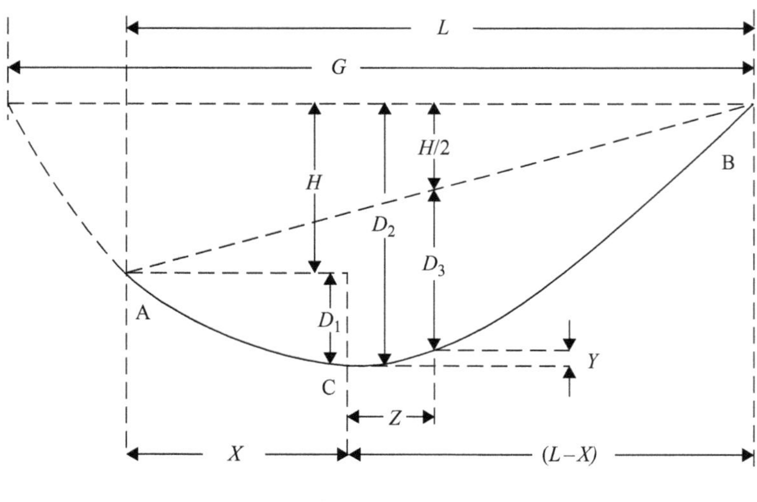

Figure 7.4 Non-level span

Slope of parabola $dy/dx = 8D_3/L^2Z$ and the slope of an inclined parabola is H/L. Therefore:

$$H/L = 8D_3/L^2 - Z \tag{7.21}$$

and, hence:

$$Z = HL/(8D_3) \tag{7.22}$$

$$Y = KZ^2 = 4D_3/L^2 \times \{HL/(8D_3)\}^2$$

$$= 4D_3/L^2 \times \{H^2L^2/(64D_3^2)\} = H^2/(16D_3) \tag{7.23}$$

Referring to Figure 7.4, the total sag on the equivalent complete span is:

$$D_2 = H/2 + D_3 + Y \tag{7.24}$$

Inserting the value of Y from equation (7.23):

$$D_2 = H/2 + D_3 + H^2/(16D_3) = H/2[1 + H/(8D_3)] + D_3 \tag{7.25}$$

If G = complete span length, then $G/2 = L/2 + Z$. Inserting the value of $Z = HL/(8D_3)$ gives:

$$G/2 = L/2 + HL/(8D_3) \tag{7.26}$$

From normal sag calculations $D_3 = WL^2/(8T)$; inserting the value of D_3 into equation (7.26) gives:

$$G/2 = L/2 + HL/8 \times 8T/(WL^2) = L/2 + HT/WL \tag{7.27}$$

$$G = L + 2TH/(WL) \tag{7.28}$$

and

$$X = L/2 - (TH/WL) \tag{7.29}$$

If the value of X is negative, the lowest point of the sag is outside the actual span. Equation (7.26) is used to establish the position in the span for the lowest sag. This is of assistance in establishing ground clearance in a hilly area, and clearances to obstacles where a line profile is not readily to hand.

7.8 Creep

7.8.1 General

Elastic increase and decrease in conductor length will, of course, be apparent as the conductor metal expands and contracts due to temperature change. However, conductors will also permanently lengthen due to non-elastic stretch, or creep.

Creep is generally acknowledged as having three components: two types of geometrical settlement due to distortion and bedding-in and a metallurgical extension.

When a conductor is stranded, the wires in alternate layers are given an opposing lay (see chapter 8) to ensure that the complete conductor retains its shape throughout its life and avoid the possibility of bird-caging. This means that each layer has only point contact with the wire(s) beneath it. Tensioning the conductor tightens the wires that tend to crush against each other. Slight deformation occurs as the conductor is tensioned such that the area of each strand reduces and the length increases until balance is achieved.

This geometrical settlement is more marked as the number of layers increases and as the diameter of the individual wires decreases. Also, the wires in each layer will settle together more intimately than during the stranding process once a load is applied. Both of these geometrical settlements will exhaust their influence over a short period of time.

There is also a permanent metallurgical extension of the conductor caused by changes in its internal molecular structure as the relative positions of metal molecules slide to take up a new arrangement, different to that at the time of manufacture, when a permanent load is applied. It is known that creep is a function of the metal used: all aluminium conductor, steel reinforced (ACSR) creep the most, all aluminium alloy (AAA) and all aluminium (AA) less and copper significantly less.

Laboratory measurements are made on actual conductors and the results achieved over a period of one to two months are extrapolated for a creep life of 10–25 years.

By temperature shift we mean that the design tension is not calculated at ambient temperature, but is shifted by a few degrees that effectively dictates that a higher tension (erection tension) is used. This temperature shift is to allow for the conductor creep (section 7.8.2).

Table 7.8 (an extract from IEC 1597 Table 5) provides mean values of temperature shift based on a creep period of ten years derived from many creep tests undertaken on stranded conductors. This table indicates suggested creep and temperature shifts.

A creep period of ten years is assumed since creep is usually small from ten to fifty years, and this level of creep may have already occurred during the stringing process.

Since there are many conductor constructions in use with many differing basic spans (and hence differing erection tensions), only a few such laboratory tests have been undertaken. As a consequence, a number of predictor equations have been

Table 7.8 Extract from IEC 1597 Table 5

Conductor	Creep (mm/km)	Mean value of temperature shift (°C)
All aluminium conductors	800	−35
All aluminium alloy conductors	500	−22
Aluminium conductor, alloy reinforced	700	−30
Aluminium conductor, steel reinforced	500	−25
Compacted covered conductor	250	−12

developed. These equations are typically of the following type for metallurgical creep:

$$\text{metallurgical creep} = K S^a H^b T^c \tag{7.30}$$

where

K = constant depending on the conductor material

S = average conductor stress (kg/mm^2)

H = time (hours)

T = temperature (°C)

a, b and c = experimentally determined coefficients

Calculations need to be made for a series of time intervals since the tension does not remain static. Computer programs have been developed which assume a time interval increment of, say, 15 per cent and the value of creep summated for the whole creep period.

Clearly, if creep is not considered during the line design, it is possible for the conductors to over-sag and infringe ground clearance resulting in an expensive operation to re-regulate the conductors to a higher tension. In view of this, a number of different creep compensation routines are employed, used either in isolation or in part combination.

7.8.2 Negative temperature shift

During laboratory analysis, it has been found that the permanent elongation of Zebra conductor is of the order of 600 mm/km (or more normally 600 microstrains). The coefficient of linear expansion for Zebra is 20×10^{-6} per °C, and this represents a change in length of 20 mm/km per °C. Hence, to fully compensate for the predicted increase in sag that the creep will produce, it is possible to erect Zebra at such a tension equivalent to a temperature shift, below ambient, of $-600/20 = -30$ °C. A similar method is used with predictor equations.

The drawback here is for a large overtension remaining for quite some period after erection, which may possibly cause vibration problems and possibly a higher tension in the conductor (and hence tower fittings) than has been allowed for in the original design. For this latter reason, NGT employs a third ruling condition for their basic sag/tension calculations, namely a fixed tension for the temperature shift employed.

In the case of Zebra, this is 40 kN at −30 °C on the assumption that stringing is unlikely to be undertaken when the actual ambient temperature is below 0 °C.

7.8.3 Positive temperature shift

Here, instead of reducing the erection tension, the temperature shift is added on to the design temperature and the sag template drawn accordingly. This allows the erection tension to be the same as the design tension at that temperature. The drawback here is the increase in costs due to the taller structure heights that will become necessary.

7.8.4 Prestressing

Since the rate of creep is large early in a conductor's life, it is possible to partially compensate for this by prestressing the conductor to some arbitrarily high tension (i.e. maximum working tension or up to 70 per cent of UTS) for a fixed period of time (one or two hours), preferably being re-regulated during that time as the tension falls back. The drawback here is the standing time required by the line gang while the conductor is being prestressed.

Once the level of creep compensation has been determined, a conductor erection table can be formulated and given to the erection gang.

7.9 Conductor clashing

Sag calculations are additionally used to calculate the minimum phase spacing between conductors to avoid conductor clash. A number of equations have been successfully used over the years, some more empirical than others. The method outlined in Engineering Technical Report ENATR 111 [4] gives greater phase spacings than earlier equations and considers the action of ice-loaded conductors as sustained by an upper and lower component of mean wind pressure (termed the gust and lull pressures). Assuming a horizontal phase separation, the gust and lull pressures have been calculated as 1.832 and 0.546 times the mean pressure, respectively. This method has been included in the software associated with ENATS 43-40 (both issues) and ENATS 43-121. This method is retained in the UK NNA BSEN 50341-3-9. Further details are included in ENATR 111 published by Energy Networks Association.

The method outlined in DIN VDE 0210 [5] for phase spacing is used quite extensively around the world. It is of the form:

$$\text{mid-span spacing} = k\sqrt{}(\text{sag}_{40} + I) + S_{\text{AM}} \tag{7.31}$$

where I is suspension insulator string length (m) (if used, otherwise 0), sag_{40} is the still air sag at 40 °C for the maximum span (m), S_{AM} is a voltage-related factor (for 11 kV, this is 0.10 m) and k is a conductor orientation and type factor (for horizontal conductors up to 150 mm^2 alloy, this is 0.7).

7.10 Additional issues to be considered

7.10.1 Uplift

As stated earlier, one of the sag/tension calculations necessary is the bare conductor sag at the cold design temperature (in the UK, −5.6 °C). The sag calculated is plotted at appropriate scales on to the plastic sag template and used to assess the presence of uplift at intermediate structures. Uplift is the action of the conductor, due to contraction, to attempt to lift vertically away from a suspension structure thereby damaging the conductor, cross-arm and/or fittings. The effect can be calculated using equation (7.28), or more simply by applying the template to the line profile and

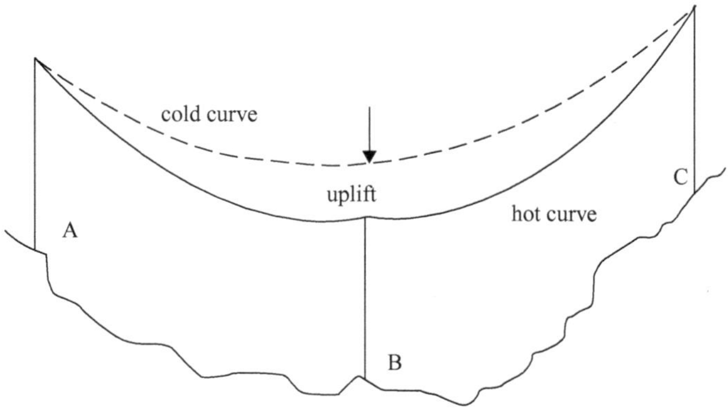

Figure 7.5 Uplift in a valley

checking visually whether uplift is a problem. In Figure 7.5 the curve is placed so that it touches structures A and C, and any uplift present at the intervening structure, B, will be instantly seen if the cold curve is above the structure.

To overcome this problem, structures A and/or C could be moved closer to B; alternatively, a taller intermediate structure at B could be used or the structure could be changed to a section (i.e. where the conductor is deadended on both sides of the cross-arm).

7.10.2 Earthwires

The calculation of sags/tensions for earthwires essentially follows that for phase wires as described above. However, it is essential that the earthwire sag does not exceed that of the phase wires, as this will increase the earthwire shield angle beyond the normally effective 30°. In extreme cases, there is also the possibility of introducing an increased risk of conductor/earthwire clash.

For these reasons, it is usual practice to specify that the earthwire design sag be matched to the design sag of the phase conductor at everyday temperature in still air. Hence, for the sag/tension calculation, the limiting criterion would additionally include the appropriate tension at everyday temperature.

7.10.3 Slack spans

The method of calculating sags and tensions as detailed above gives a good deal of correlation in practice. Occasionally, however, it is necessary to give special consideration to situations where conductors are required to be erected slack.

Sub-station gantries are often made of reinforced concrete, but will not permit high conductor tensions. In these situations, it is usual to erect conductors as a short, low-tension, single-span section (30–50 m) to relieve the gantry loading.

Due to the shortness of the span, the tension insulator set length and weight, particularly for higher voltage lines, has a material bearing on the final sag and tension produced, and it is important to accurately calculate both to ensure that adequate clearances are maintained and the gantry is not overstressed. One formula that has been used successfully is a variation of the basic sag/tension formula of equation (7.12):

$$T_2^2[T_2 + \{W_1^2 + IM/L^2(12W_1^2 + 8M/L)\}L^2Ea/(24T_1^2) - T_1 \pm Ea\alpha(t_2 - t_1)]$$
$$= \{W_2^2 + IM/L^2(12W_2^2 + 8M/L)\}L^2Ea/24 \quad (7.32)$$

where I is insulator string length (m) and M is insulator string mass (kg) and the sag is:

$$D = [(W_2L^2) + 4IM]/(8T_2) \quad (7.33)$$

Example

Consider a tension set 1.5 m long with a mass of 50 kg (e.g. 132 kV) mounted at a terminal structure and sub-station gantry. The span length is 35 m and conductor 200 mm^2 AAAC Poplar. The maximum working tension for the downlead is normally limited to 4.44 kN (1000 lbs).

Now, consider two scenarios, one with and one without the tension insulator string included in the calculation to compare the different loadings, assuming that the design tension at 5 °C is the same in both cases; see Table 7.9.

Although the MWT value has reduced when the effect of the insulator string is included in the calculation, the design sag (in this case assumed to be 50 °C) has increased, and clearly this could be crucial when assessing ground/obstacle clearances.

7.10.4 Short single-span sections

From practical experience, Balfour Beatty have found that to string a conductor satisfactorily, the sag during erection must be not less than indicated in the following empirical formula [6]:

$$\text{erection sag (m)} >= \text{span length (m)}/60 \quad (7.34)$$

Table 7.9 Slack span sags and tensions

	Slack span section 35 m			
	Normal parabolic tension (kgf)	Sag (m)	Tension amended for insulator string (kgf)	Sag (m)
Max. working tension	452.8		347.4	
Temperature 5 °C	144.1	0.72	144.1	0.98
Temperature 50 °C	105.0	0.99	127.5	1.11

If normal sag/tension limiting criteria were applied to short single-span sections, the difference between the actual true length of the conductor and the physical span length would be so small as to prohibit landing of the conductor.

Reducing the stringing tension to land the conductor does not usually introduce ground clearance issues (as sags remain small), although for wood pole lines counter-balance stays would be necessary to cater for the differential tension at the section structure. Such a situation is often not acceptable to landowners, and a compromise becomes necessary where the short single span is included temporarily in an adjacent section during stringing. The short section is then cut-in once the elongated section has been strung. It is important to note that the inclusion of a short section as described would likely have a serious impact on the actual equivalent span for the section and should only be considered once sags/tensions have been re-calculated for the temporarily longer section, and a full assessment of structure strength made which considers the differential load during worst loadings.

7.10.5 Effect of steep hills on conductor tensions

We have seen earlier that the total tension in a span is greater than the horizontal tension calculated by traditional methods due to the effect of resolved vertical loads. For spans in reasonably flat areas this difference is very small, however, hilly areas introduce greater downpull such that this effect cannot be ignored.

When any line is strung, the total tension in any span will be the same throughout the section, assuming there is little friction in the conductor rollers. As the ground profile becomes hillier, so the vertical component of tension increases, or decreases, depending on whether the line travels uphill or downhill, respectively. This means that spans at the top of a hill will have less sag, and, hence, more ground clearance than planned and those at the bottom more sag and less ground clearance.

Clearly, those spans at the bottom of a hill will need special consideration to ensure that statutory ground clearance has not been eroded. To assess the effect of the ground profile on conductor sag and tensions, it is necessary to calculate the horizontal tensions in each span as follows:

$$T_H = T_{H'} \pm WH$$

where H is the height difference between adjacent structures, assuming the sags in each span are reasonably similar.

For lines with suspension strings at intermediate structures, the calculated value of T_H becomes $T_{H'}$ for the next span, and so on, and hence the effect on horizontal tension becomes cumulative as the gradient continues.

For the reasons stated above, where sagging boards are used to check the sag during erection (i.e. 132 kV portal construction), the check span should be somewhere near the centre of the line section. The tension should then settle down, after the suspension sets are plumbed, to something fairly near the correct horizontal tension.

Lines employing pin or post insulators will have the unequal tensions in a hilly section built in. To overcome the problems of overtensioning the upper spans of a hilly section, the line should be sagged using sagging boards (not a dynamometer) on the top-most span. This will now mean there is no overtension in the whole section, however, the lower spans will now be under tension, this situation being remedied at the profiling stage by allowing a greater ground clearance in these spans.

7.11 Summary

This chapter has shown how to calculate the sag and tension requirements for spans on level and hilly ground. This is essential so that electricity regulations are met at every stage in the lifetime of a line. The effect of creep has also been discussed and the concept of a negative temperature shift in the erection tension has been explained. Wind damage that can be caused by conductor clashing is allowed for by the phase spacing, and, finally, the serious problems of uplift in cold conditions have been discussed.

7.12 Appendix A – weight of ice-loaded conductor

Assume a conductor has a bare diameter of d (mm), and an ice accretion thickness of R (mm) (Figure 7.6). Projected diameter of the ice-loaded conductor is:

$$d + 2R$$

or

$$2(r + R) \text{ (mm)}$$

Area of conductor is:

$$\pi r^2 \text{ (mm}^2\text{)}$$

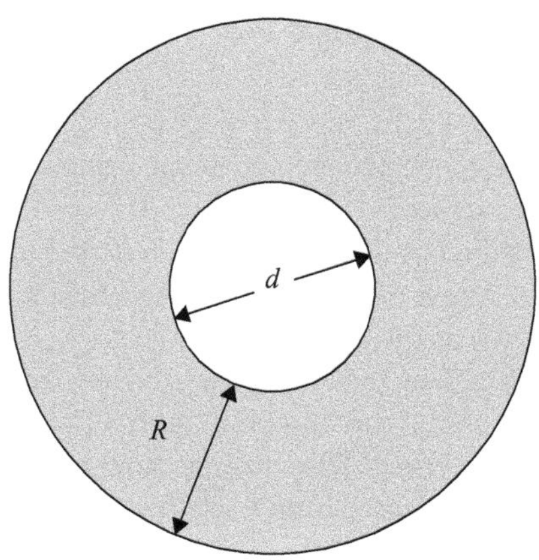

Figure 7.6 Schematic of ice-covered conductor

Area of ice-loaded conductor is:

$$\pi(r + R)^2 = \pi r^2 + \pi R^2 + \pi(2rR)\ (\text{mm}^2)$$

Hence, sectional area of ice only is:

$$\pi R^2 + \pi(2rR)$$

This reduces to:

$$\pi R(R + d)\ (\text{mm}^2)$$

Now, density of non-aerated ice approximates to 0.913 g/cm^3 and hence weight of ice is:

$$0.913\pi R(R + d)/100 \text{ g/cm run of conductor}$$

In terms of an augmented diameter of ice (as defined in OHL Regulations 1970), it is:

$$0.913\pi/4 \times 2R(2R + 2d)\ (\text{g/m})$$

or

$$0.717 \times 2R \times (2R + 2d)\ (\text{g/m})$$

Adding the bare weight of the conductor, the total ice-loaded weight becomes:

$$W_B + K/1000 \times 2R \times (2R + 2d)\ (\text{kg/m}) \qquad (7.\text{A}1)$$

The factor K will vary depending on the ice density employed as follows:

glaze ice (0.913 g/cm^3): $K = 0.717$

wet snow (0.850 g/cm^3): $K = 0.668$

rime ice (0.510 g/cm^3): $K = 0.401$

7.13 References

1 Cigré SCB2 WG11 TF04: 'Safe design tension with respect to Aeolian vibration', *Electra*, 2001, **198**, pp. 28–43
2 prIEC 61989: 'Overhead electrical conductors – guide to calculation of sag and tension', committee draft 7/519/CD (12 March 1999)
3 BRADBURY, J., KUSKA, G. F., and TARR, D. J.: 'Sag and tension calculations for mountainous terrain', *IEE Proc.*, 1982, **129**(5)
4 ENATR 111: Engineering technical report, Energy Networks Association, London 1989
5 DIN VDE 0210: 'Conductor clashing calculations for bare wire overhead lines'
6 ORAWSKI, BRADBURY and VANNER: 'Overhead distribution lines – some reflections on design', *IEE Proc. C*, 1986, **133**(7)

7.14 Further reading

IEC 1597: 'Evaluation of conductor creep and the use of temperature shifts'

MINER, M. A.: 'Cumulative damage due to fatigue', *Proc. ASME J. Appl. Mech.*, 1945

Cigré SC22 WG04: 'Recommendations for the lifetime evaluation of transmission line conductors', *Electra*, 1979, **63**, pp. 103–145

ZETTERHOLM, O. D.: 'Mechanical calculations on overhead conductors', Cigré session, 1960, report no. 223

Chapter 8

Conductor characteristics and selection

8.1 Scope

This chapter considers the geometry of conductors and the various materials used. This is followed by a look at the electrical requirements, how to calculate them and a look at the current ratings and greasing (corrosion) requirements of bare and covered conductors. The covering can be simply a thin sheath or full insulation. In the former, the conductor is covered to reduce clashing problems or for safety or space reasons but it is not insulated. It must be remembered, however, that the best choice electrically may not be the best choice mechanically or even environmentally.

Conductor selection is a complex business. It is dependent not only on electrical, magnetic and economic considerations, but also on radio interference, stray fields and fault level capability. Mechanically, the conductor must support its own weight as well as any ice and wind loading, and must maintain ground clearances under all predicted environmental conditions.

It is hoped that this chapter will familiarise the reader with the characteristics of a range of commonly used conductors and list the criteria for conductor selection for a particular job. As an aide for information, a section is included giving a brief history of the development of covered conductors.

8.2 Conductor characteristics

The range of physical characteristics to be considered as far as overhead line (OHL) conductors go are listed in Table 8.1.

The characteristics required can be obtained by a combination of conductor geometry, material selection and control of the manufacturing process. It will be shown how these can be manipulated so that various options become available and an informed choice can be made.

Table 8.1 Conductor characteristics

Property	Reason
High conductivity	to give adequate current carrying capacity and low voltage drop
High strength	to maintain ground clearance in long spans
Low weight	as above
Flexibility	avoid vibrational fatigue failure
Mechanical stability	to withstand a variety of loading conditions
Physical stability	to withstand environmental conditions (e.g. corrosion resistance)
Lifetime stability	to maintain the characteristics for 40–50 years

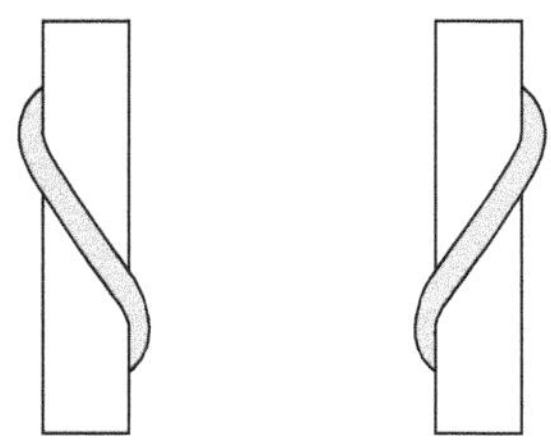

Figure 8.1 Twist directions

8.3 Conductor geometry

Although there are many different types of conductor geometry, including segmented and oval sections, only the conductors made from standard circular cross-section strands will be covered in this section. Novel developments will be looked at later.

Conductors are generally made up of individual strands in order to improve flexibility and conductivity. In AC current the higher the frequency, the less the current actually penetrates into the surface of the strand. At MHz frequencies (as with a lightning surge) the penetration (or skin depth as it is known) will only be a few tens of microns. Due to this skin effect most of the current at 50 Hz is carried in a surface layer around 1 mm thick. So for strands of, say, 3 mm diameter, almost all the conductor section will be available for current carrying capacity. Extra large conductor strands will thus add weight but give little increase in current carrying capacity. A solid conductor would carry even less current and would also suffer from vibration fatigue. A further gain in flexibility is obtained by reversing the twist in alternate layers. By convention, the outer layer often has a right-hand twist (Figure 8.1).

The standard geometry for strands of the same size means that there will be a central strand followed by a layer of 6 strands, then 12, then 18, 24 etc. So conductors will normally have 7, 19, 37 or 61 strands etc. Some conductors,

Table 8.2 Conductor specifications

Codename	Type	Nominal area (mm^2)	Actual area aluminium (mm^2)	Stranding number/diam (mm) aluminium	Stranding number/diam (mm) steel	Weight (kg/km)
Ant	AAC	50	52.83	7/3.10		145
Hazel	AAAC	50	59.87	7/3.30		164
Rabbit	ACSR	50	52.88	6/3.35	1/3.35	216
Horse	ACSR	70	73.36	6/3.35	1/3.35	546
Wasp	AAC	100	106.0	7/4.39		290
Oak	AAAC	100	118.9	7/4.65		325
Dog	ACSR	100	105.0	6/4.72	7/1.57	396
Caterpillar	AAC	175	185.9	19/3.53		511
Elm	AAAC	175	211.0	19/3.76		659
Lynx	ACSR	175	183.4	30/2.79	7/2.79	865
Caracal	ACSR	175	184.2	18/3.61	1/3.61	600
Butterfly	AAC	300	322.7	19/4.65		888
Upas	AAAC	300	362.1	37/3.53		997
Goat	ACSR	300	324.3	30/3.71	7/3.71	1530

especially those of dissimilar materials, may have a larger central strand or even a multi-stranded core.

There are other versions such as 32 mm^2 hard drawn copper that has only three strands. In a manufacturer's conductor catalogue the following descriptions for some aluminium conductor steel reinforced (ACSR), all aluminium alloy (AAAC) and all aluminium (AAC) conductors will be given. Note the considerable differences in mechanical characteristics for the same nominal sizes (Table 8.2).

Horse, an ACSR, has 19 strands each of 2.79 mm diameter, the central seven forming the steel core. Lynx is similar but larger with 37 strands. However, Caracal is the same size and conductor type (ACSR) as Lynx, but is made up of only 19 strands. Finally, Dog has a central steel core of seven small steel strands and just one layer of aluminium. All these variations give the line designer many options. By UK convention all ACSR conductors are given animal names. All aluminium conductors (AAC) are named after insects and all aluminium alloy conductors (AAAC) after trees and shrubs. With the recent publication of BS EN 50182, conductor names are changing to a numerical code based on the aluminium and steel areas and the aluminium grade. For example, 100 mm^2 Dog ACSR will now be identified as 105-AL1/14-ST1A and the same format is beginning to appear in other countries around Europe.

In the mechanical construction of the conductor there are two important parameters, the lay length and lay ratio. The lay length is the distance measured along the conductor axis in which any one strand makes one complete revolution. The lay ratio is the ratio of the lay length to the external diameter of that layer. In order to avoid birdcaging, where the inner layers bulge out of the conductor under stress, the lay ratio should be less than that of the layer beneath it. This rule is normally satisfied, as the inner layer will have a smaller diameter anyway.

There are instances where the conductor diameter may need to be reduced, which can be achieved by compacting it during manufacture. The process deforms the strands and eliminates the interstices in between. This reduces wind loading but also decreases flexibility. Compacted conductors are often named after towns (e.g. Blyth).

8.4 Material selection

8.4.1 General

In order to obtain the desired conductivity and strength the first port of call is the material to be used. This may not be suitable for all our requirements (e.g. the best material for strength and conductivity may corrode too quickly) but heat treatment could help the situation.

The most common materials used for OHL conductors are:

- copper – hard drawn (HD)
- cadmium copper alloy
- aluminium
- aluminium–magnesium–silicon alloy
- galvanised steel
- aluminium clad steel.

Cadmium copper is now banned from use for new lines in Europe, but it is included for completeness as there are many existing lines using this material.

The conductor characteristics are summarised in Table 8.3.

The treatment of the material during manufacture can introduce specific characteristics, e.g. the process of drawing down the feed rod introduces an increase in strength with a slight loss in conductivity. If this is not annealed out, then the material is known as hard drawn (HD).

Steel wire can be used as a strength member but must be protected against corrosion. This is accomplished by coating with zinc (galvanising) or aluminium (by a continuous extrusion process). As the steel condition is important to the safety of the line, the integrity of the zinc coating is essential. This can be checked on line by the use of a high-frequency pulse. The pulse frequency is chosen so that the induced current will only penetrate the steel roughly equivalent to the thickness of the zinc layer due to the skin effect. The presence, or not, of the zinc will give a substantially different response.

So, an OHL conductor can be made from a single material or a combination to suit. The choice is not always simple. For instance, copper has three times the density of aluminium but only 50 per cent higher conductivity, and so will give a heavier conductor for the same current carrying capacity. However, copper is far

Table 8.3 Conductor characteristics

Parameter	Units	Hard drawn copper	Cadmium copper	Hard drawn aluminium	Aluminium alloy	Galvanised steel	Aluminium clad steel
Conductivity	% IACS[1]	97	79	61	53	9	20
Resistance	$\Omega\,mm^2/km$	17.7	21.8	28.3	32.5	192	84.8
Temp coeff. of resistance	per °C	0.0038	0.0031	0.0040	0.0036	0.0054	0.0051
Coeff. of linear expansion	$\times 10^{-6}$ per °C	17	17	23	23	11.5	13
Linear mass	$kg/mm^2\,km$	8.89	8.945	2.7	2.7	7.8	6.59
Ultimate tensile stress (UTS)	N/mm^2	414	621	160–200	295	1320–1700	1100–1344
Modulus of elasticity	Gpa	125	125	70	70	200	162

[1] % IACS is the percentage conductivity of the material in relation to pure copper.

more corrosion resistant than aluminium and will have a longer life in coastal areas. It can also be erected at a higher tension – so reducing sags and increasing ground clearance. But yet again, aluminium is stronger than copper for the same weight – reducing pole loading. The options are wide and no conductor is perfect. However, a brief description of the more common conductor types is given next.

8.4.2 Common types of conductor

8.4.2.1 AAC – all aluminium conductor

As the name implies, all aluminium conductor is made of electrical grade purity hard drawn aluminium strands throughout. This is a low-cost conductor, but with a limited strength-to-weight ratio. This can restrict the span lengths that can be used in areas where wind and ice loads can be high. BS EN 50183 recognises this material as AL1 (i.e. almost pure aluminium).

8.4.2.2 AAAC – all aluminium alloy conductor

All aluminium alloy conductor is a heat-treated alloy that achieves a higher strength than AAC. It has a superior strength/weight ratio to ACSR and is harder than AAC, and therefore less susceptible to surface damage. It can be used to reduce sags or increase span lengths compared with AAC. It also has a greater strength-to-weight ratio than ACSR for larger sizes and generally a lower AC resistance (lower losses) and no galvanic corrosion. Salt corrosion can still be a problem in marine environments. A number of alloy grades are available according to BS EN 50183 and these are indicated as grades AL2–AL8. The differing alloy grades trade conductivity for strength, as indicated in Table 8.4.

Table 8.4 Aluminium and alloy grades to BS EN 50183

Type	Nominal conductivity (% IACS)	Nominal diameter (mm)	Minimum tensile strength (N/mm^2)	Min. elongation after fracture on 250 mm (%)	Maximum resistivity (nΩ-m)
1	61.0	1.50–1.75	190		28.264
	61.0	1.75–2.00	185		28.264
	61.0	2.00–2.25	180		28.264
	61.0	2.25–2.50	175		28.264
	61.0	2.50–3.00	170		28.264
	61.0	3.00–3.50	165		28.264
	61.0	3.50–5.00	160		28.264
2	52.5	1.50–3.50	325	3.0	32.84
	52.5	3.50–5.00	315	3.0	32.84
3	**53.0**	**1.50–5.00**	**295**	**3.5**	**32.53**
4	52.9	1.50–3.50	325	3.0	32.9
	52.9	3.50–5.00	315	3.0	32.9
5	55.25	1.50–5.00	295	3.5	32.2
6	55.6	1.50–3.50	314	3.5	31.5
	55.6	3.50–5.00	304	3.5	31.5
7	57.5	1.50–2.50	300	3.0	30.5
	57.5	2.50–3.00	290	3.0	30.5
	57.5	3.00–3.50	275	3.0	30.5
	57.5	3.50–4.00	265	3.0	30.5
	57.5	4.00–5.00	255	3.0	30.5
8	57.5	1.50–2.50	300	1.6	30.5
	57.5	2.50–3.00	290	1.8	30.5
	57.5	3.00–3.50	275	2.0	30.5
	57.5	3.50–4.00	265	2.4	30.5
	57.5	4.00–4.50	255	2.7	30.5
	57.5	4.50–5.00	245	3.0	30.5

Alloy grade AL3 is indicated in bold in this table, as this is the alloy that has been used historically in the UK for all standard aluminium alloy conductors.

8.4.2.3 ACSR – aluminium conductor steel reinforced

A method of significantly increasing the strength of AAC is to have a galvanised steel core. The mechanical performance is also improved by its higher modulus of elasticity and lower coefficient of expansion. Although more expensive per metre than AAC, its greater strength (and hence longer span lengths – fewer poles) means that the overall line cost can be lower. This makes it very popular for medium voltage lines. The steel core may be a single strand or be multi-stranded. If the steel is aluminium clad instead of galvanised then there is a conductivity gain but some loss of other mechanical properties. This type is known as ACSAR (aluminium conductor steel/aluminium reinforced) and is relatively expensive compared with ACSR.

8.4.2.4 AACSR – aluminium alloy conductor steel reinforced

Another ploy to increase strength in circumstances where conductivity is less important is to replace the aluminium in ACSR with aluminium alloy. Earthwire applications are one area where this could be used.

8.4.2.5 ACAR – aluminium conductor alloy reinforced

Aluminium conductor alloy reinforced will not suffer from galvanic corrosion (no steel) but will have increased conductivity over AAC but lower strength than ACSR.

8.4.2.6 HD copper – hard drawn copper

The most common conductor used within a few kilometres of the coast purely for its high corrosion resistance to salt is hard drawn copper.

8.4.2.7 Cadmium copper

A copper alloy also used for its corrosion resistance and hardness, but now no longer used due to environmental considerations. There are many lines in the UK strung with this conductor, which was eminently suitable for long span situations.

8.4.2.8 Compacted conductors

ACSR and AAAC conductors can be compacted by a final die drawing process to give an overall reduced diameter without reducing current capacity. This can reduce wind and ice loads but can reduce flexibility. Another advantage is the ability to increase the steel core size (e.g. by 200 per cent), thereby substantially increasing strength-to-weight ratios without increasing wind loads.

8.4.2.9 Covered conductors

Covered conductors will be discussed in detail in chapter 9. Essentially, covering the bare metal conductor with an insulating or insulated sheath reduces clashing problems, is wildlife friendly and can be far safer when touched accidentally in leisure activities or machinery. Corrosion is also reduced but particular attention has to be given to lightning protection and vibration.

8.4.3 Specific scenarios

To appreciate the fact that different conductor characteristics are needed for different scenarios, consider the following examples.

8.4.3.1 A medium voltage line leaving the sub-station

The first stretch of line leaving a sub-station will carry all the current that the network section will use. It will therefore need a high current capacity and security, as failure would put many customers off power supply. So there will be a need to build in sufficient spare capacity to allow for faults and emergencies and possibly future expansion.

8.4.3.2 10 km long medium voltage spur line feeding just 20 customers

The long, underused spur line will suffer voltage drop when power is used – thereby causing lights to dim. However, to put heavy conductor up is expensive, and, having so few customers, it may also not be a very important line. So a conductor is needed that avoids volt drop but is also inexpensive.

8.4.3.3 A line across a river or gorge

A line across a river or gorge may need to be a very long span or alternatively be supported in water. This is done on some occasions in harbours or lakes. The alternative is to build strong supports and use high strength/weight ratio conductors.

8.4.3.4 Lines on Dartmoor, in the Peak District and in the north of Scotland

The environmental effects of the area on the conductor have to be considered. Dartmoor is on a peninsula (Cornwall) and so suffers severe salt corrosion as well as low cloud and mists (damp) leading to ice in winter. The conductor needs to withstand these conditions. The Peak District does not suffer from any salt effects but may suffer snow loads and high winds, relatively few customers, isolated lines difficult to access in the case of faults. So consider reliability as a major factor as well as low electrical but high mechanical loads. The north of Scotland will suffer severe blizzards, isolated areas with few customers but also salt corrosion near the coasts. So the conductor needs to withstand high mechanical loads but low electrical loads and must also withstand salt corrosion. Maybe light conductors and short spans are the answer. What about building in weak points for those blizzard conditions?

To go back briefly to scenario 8.4.3.3, in the UK there is little need for really long spans, the Severn crossing perhaps being the longest. Consider the problems of Norway with many lines crossing fjords, but even that country does not boast the current record for long spans. This is held by Greenland with a span of 5 km – just over three miles.

8.5 Conductor life

8.5.1 Causes of failure

There are many factors that will cause conductors to fail in service. The preponderance of one or other process depends on the way the conductor is used and the environment in which it is placed. The main causes of premature failure are:

- fatigue
- creep
- corrosion.

8.5.2 Fatigue

Fatigue is caused by the continual movement (flexing) of conductors at a fixed point where they are held. It is commonly due to the vibration that is caused by wind blowing across the conductor (Aeolian vibration). Each conductor has its tension limit above which fatigue becomes a problem, and this limits the span lengths unless antivibration methods are used. Typically, aluminium conductors can be strung to 20 per cent of their UTS and copper conductors to 33 per cent of their UTS. Cigré reports [1–3] give full details. Vibration can cause conductor failure at connectors or clamps due to the dynamic bending stresses (the clamping pressure restricting strand movement) and any initiating surface damage. The vibration is caused mainly by wind speeds in the range 2 to 20 mph with most damage caused in light winds. Vibration is also dependent on the line rigidity, i.e. it increases with line tension and so is also temperature dependent as line tension increases as the temperature drops. Aeolian vibration occurs in the frequency range from 3 to 100 Hz.

A less frequent wind-induced problem is galloping, where the conductor oscillates at low frequency (<3 Hz) and high amplitudes (up to the sag which can be several metres on transmission lines). This is more common on transmission tower lines. Galloping occurs under relatively high wind speeds (above 15 mph). Aerodynamic forces are coupled through the angle of attack (wind direction relative to the conductor). There may also be mechanical coupling when natural frequencies of different motions coincide. Typically, this happens when eccentric ice weight and aerodynamic torsional moment reduce the torsional natural frequency until it is equal to the vertical natural frequency. The conductor becomes aerodynamically unstable – normally due to an asymmetric rime ice deposit. The conductor orbit is usually an ellipse whose major axis is vertical and minor axis horizontal.

Finally, sub-span oscillations can occur between the spacers of bundled conductors. They are a function of conductor diameter, spacing, wind angle and wind speed, and generally have the same frequency as galloping but with amplitudes up to the conductor spacing only.

8.5.3 Creep

Creep is the inelastic extension of a material under stress, and has been covered in detail in chapter 7. The creep can be calculated from laboratory experiments carried out under extremely strict conditions of temperature control – usually in long underground rooms where a 60-m span can be maintained under disturbance-free conditions.

8.5.4 Corrosion

Corrosion is the normal life-limiting factor in the UK. This is in contrast to many other parts of the world. It appears that our environment is a particularly aggressive one with respect to aluminium conductors, and the problems associated with corrosion can be particularly acute. Aluminium is generally considered to be a corrosion resistant material, however, thermodynamically it is a very reactive material and relies on the formation of a very stable oxide layer on its surface to prevent corrosion in normal circumstances. However, as for any material that relies on a protective coating to prevent corrosion, situations that give rise to damage or prevent formation of this oxide coating can result in rapid corrosion. Two particular mechanisms are relevant to aluminium used in stranded conductors. First, crevice corrosion. In an area of restricted oxygen supply, such as the bottom of a crevice, the maintenance of a coherent oxide fill can be compromised and this leads to a differential aeration cell set up between the base of the crevice and the mouth, where more oxygen will be available. The result of this will be pitting corrosion at the base of the crevice. The other issue of particular relevance is the action of aggressive anions such as chloride ions. As a result of their size and charge density, such ions are particularly effective at infiltrating and disrupting the oxide film and therefore increase the risks of corrosion damage at susceptible points such as the base of crevices.

Based on this brief description of corrosion processes it is clear that the design of overhead conductors has created a very good environment for corrosion. If moisture and pollution penetrate into the interstices of a stranded conductor, corrosion will occur. If the pollution contains significant chloride ions, i.e. marine pollution, this can result in severe corrosion in relatively short periods of time.

The lifetime of bare aluminium conductors in the UK can vary from less than five years to greater than 50 years. In extreme situations, with small conductors that were ungreased or poorly greased at the time of manufacture, failures have occurred in aggressive coastal environments in less than two years. Larger conductors with several layers of aluminium and with larger cross-section individual strands, which are well greased, will have lifetimes in excess of 50 years in relatively benign inland environments.

The damage that ultimately causes failures to conductors occurs as the result of internal corrosion. That is, corrosion due to the mechanisms described above when pollution and moisture penetrate into the interstices of the conductor. In order to accommodate the larger corrosion product, the strands need to deform. Once they have significant pitting corrosion they are unable to deform and fracture will occur. When strands fracture there is a transfer of current between strands which leads to accelerated corrosion due to the effects of AC which are particularly significant for aluminium. This process, therefore, is an accelerating process that ultimately causes the mechanical failure of the conductor.

Grease is used internally in conductors to prevent the ingress of moisture and pollutants in order to prevent the corrosion process occurring and therefore to increase the lifetime of conductors significantly. This grease is applied at the time of manufacture, and if applied effectively can lead to extended lifetimes even in relatively aggressive conditions. It should be noted that the grease itself does deteriorate when exposed to severe marine pollution, particularly for small conductors, and therefore even well greased conductors do have a limited life in aggressive conditions. The level of greasing, the consistency of its distribution and its condition are therefore additional factors affecting conductor lifetime.

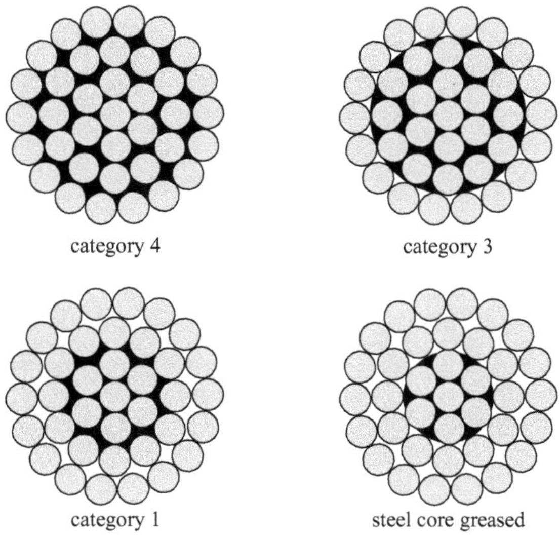

Figure 8.2 Greasing categories

Greases fall into two main categories – hot-applied and cold-applied:

1 Hot-applied greases are essentially oil within a wax matrix. The drop point or melting point is set to be above the expected maximum temperature the conductor will reach.
2 Cold-applied greases are essentially synthetic oils that may decompose before reaching their drop point. These are easier to apply but may not be suitable for high-temperature operations.

Greasing is also applied to different levels according to the corrosion risk – from category 4 (all strands greased) to the steel core only being greased. Figure 8.2 shows the categories schematically. They are detailed in ENA Engineering Recommendation L38/1 [4].

8.6 Manufacturing the conductor

This section is essentially for background information. Engineers may at times need to be satisfied that a particular manufacturer has high-quality production techniques, and so may feel justified in seeing how a company makes its conductor. Is the process clean, efficient, reliable? what are the quality control standards etc.? In order to grasp what is happening, the basics of conductor manufacturing can therefore be a useful starting point.

Basically, the manufacturing process consists of drawing the metal down from the delivered rod size to the required wire size and then stranding it to form the conductor. This is fine for most basic materials. However, the processing of aluminium

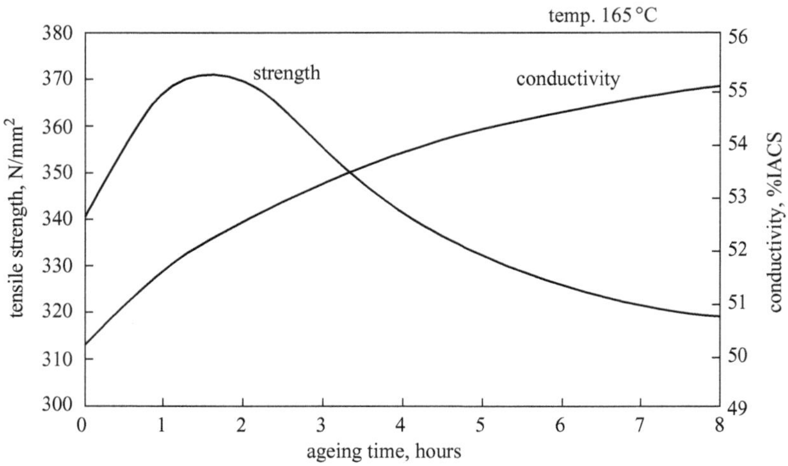

Figure 8.3 Heat treatment schematic for aluminium alloy wire

alloy is more complex and offers the manufacturer a further tool in his armoury of manufacturing specific characteristics.

For an AAAC, the initial rod is manufactured from various alloying elements such as magnesium, silicon, iron and copper dissolved into the base aluminium. As the material is an alloy and not a compound, the various elements will diffuse out with time and temperature, thereby changing the characteristics of the material. To avoid this, the rod is drawn down (which work-hardens the material) and then heat treated to allow the alloying elements to change the characteristics until the required values are achieved. On a simple basis, the strength and conductivity will vary according to the temperature and time of heat treatment according to the graph in Figure 8.3.

Stranding sounds easy – but it is not that straightforward. First of all the centre core is made up – either as a single wire or as stranded core in the case of ACSR. Greasing is nearly always applied here. Then, for each strand in the next layer, a rotating drum is maintained around the central core of the stranding machine – like planets orbiting the sun. This analogy results in this machine being known as a planetary strander. Each successive layer is applied with the planets of drums rotating in opposite directions (Figure 8.4). Grease may be applied with each layer as desired. However, this is not all. If it were, the conductor would spring apart as soon as it was cut. In order to avoid this, the individual strands are either set by preforming immediately before the stranding die or the complete conductor is worked by post-forming on a capstan after stranding.

After the conductor is made it has to pass various type and sample tests as specified in the relevant national and/or international standards.

8.7 Electrical design considerations

This section will deal with the key electrical design considerations necessary to select the appropriate conductor for a given application. These are the continuous

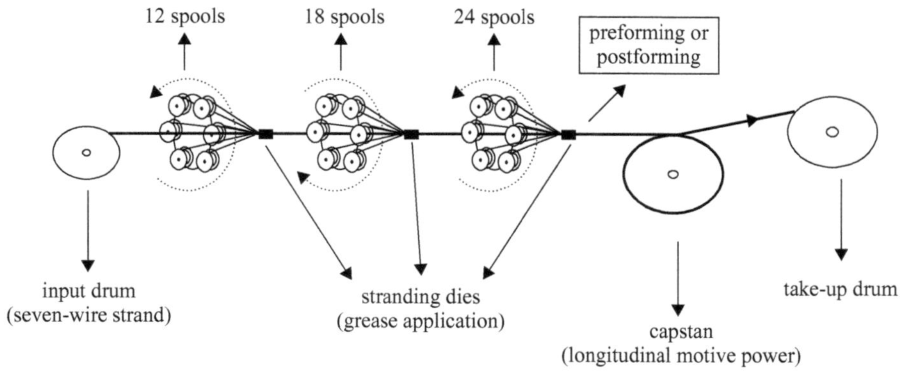

Figure 8.4 Schematic of a planetary stranding machine

and short-time current rating, inductive and capacitative reactance and voltage drop, which are only some of the factors that are required. Other considerations to be taken into account may include cost, mechanical suitability, standardisation for stock and maintenance simplification etc.

8.7.1 Technical requirements

8.7.1.1 Current rating

Current rating is of primary importance. The conductor, when carrying current, will lose energy through resistive losses and gain energy from solar gain. There will be counterbalancing radiative and convective losses from the conductor surface. High conductor temperatures can be used to increase current carrying capacity but this can lead to loss of strength, increased sag and increased oxidation (corrosion).

8.7.1.2 Resistance

DC resistance is dependent only on the conductor temperature. The various standards specify strand resistance and the conductor resistance is calculated from that.

AC resistance is dependent on the skin effect and is therefore dependent on current frequency and the material used. On steel-cored conductors it is also dependent on the magnetic properties of the core.

8.7.1.3 Reactance

Inductive and capacitive reactances are present in AC lines. They are dependent on conductor diameter, phase spacing, line height and length, frequency and current levels. Reactance is therefore a function of the total line design.

8.7.1.4 Corona

High voltage levels on a conductor will cause local ionisation of the air and cause energy loss, noise (audible crackling) and radio interference. The voltage threshold at which this occurs depends on the conductor diameter and surface roughness as well as air temperature, humidity and pressure. It can be critical in lightning protection and long distance energy losses on transmission lines – especially in winter.

8.7.1.5 Fault currents

Faults caused by network problems or lightning strikes or deliberate vandalism can cause high fault currents in the conductors. The heating effect on the conductor limits the fault current it can carry for any specified time. This is usually quoted as I^2t where I is the fault current in kA and t is the time in seconds.

8.7.1.6 Sag and tension

Sag and tension calculations are very important as the line has to meet statutory safety and electrical clearances. These can be readily calculated from the conductor characteristics and expected temperature in service. The modulus of elasticity and the coefficient of linear expansion are both important parameters in these calculations.

8.7.2 Continuous current rating for bare conductors

The current rating depends on the parameters given in Table 8.5.

The calculation of the current rating is based on the fact that when equilibrium is reached at the maximum operating temperature, T_m, heat balance is achieved where heat input is equal to heat output.

A conductor gains heat due to its electrical resistance and solar gain. It loses heat by wind-assisted convection and by radiation. In 1949 the British Electrical and Allied Industries Research Association published the following empirical formulae for the convection loss and the radiation gains and losses.

8.7.2.1 Convection

$$\text{convection loss} = 387.1(T_m - T_a)(Vd)^{0.448}\ \text{W/m} \tag{8.1}$$

where d is the conductor diameter in mm and V is the wind speed in m/s.

The wind speed (V) is strictly the effective speed, which is the actual speed, corrected for temperature and pressure – but in most cases the actual speed may be used with little error.

Table 8.5 Parameters for current rating

Parameter	Symbol	Unit
Design operating temperature of the line	T_m	°C
Ambient temperature	T_a	°C
Intensity of solar radiation	S	W/m^2
Wind speed	V	m/s

Table 8.6 Emissivities of common conductor types

Conductor	Condition	
	new	old
AAC, AAAC, ACSR	0.3	0.9
Copper, cadmium–copper	0.2	0.9

There are different international standards, but in the UK a speed of 0.447 m/s (1 mph) is normally assumed.

8.7.2.2 Radiation

$$\text{radiation loss} = 1.791 \times 10^7 \times \varepsilon d\{(T_m + 273)^4 - (T_a + 273)^4\}\ \text{W/m} \qquad (8.2)$$

where ε is the emissivity of the conductor (no units).

This is a problem for aluminium. When new and unoxidised it has a low emissivity of around 0.2. Within a short time, depending on the atmosphere, it weathers to an emissivity of above 0.5. Finally, and for most of its life, the conductor will have an emissivity of nearly 0.9. This is the value normally used. Recommended values for the different types of conductor are given in Table 8.6.

$$\text{solar energy gain} = Y \times S \times d\ \text{W/m} \qquad (8.3)$$

where Y is the solar absorption coefficient (no units). Again, the value for the absorption coefficient depends on the condition of the conductor – values of 0.6 (new) and 0.9 (old) are recommended for all conductor types. The UK standard for solar radiation is 580 W/m^2.

8.7.2.3 Electrical resistance

$$\text{electrical energy gain} = I^2 R_{20}[1 + \alpha(T_m - 20)] \qquad (8.4)$$

where:

I = current (A)

R_{20} = resistance (at 20 °C) of the conductor (Ω/km)

α = temperature coefficient of resistance at 20 °C (/°C)

R_{20} is normally taken as the DC resistance. Using the DC value gives less than one per cent error for conductors (<250 mm^2 area) containing no steel. For an AC rating, the value of R_{20} that is used should be the AC resistance. A similar R_{AC}/R_{DC} ratio is associated with ACSR when there is an even number of layers of aluminium wires. If the number of aluminium layers is odd, however, the effect of the magnetic flux in the core, which varies according to the load current, gives rise to increased

R_{ac}/R_{dc} ratio values. The effect is moderate in three-aluminium-layer conductors and considerable in one-aluminium-layer conductors (up to approximately 20 per cent for high current loads). R_{ac} can be calculated using more accurate methods [5, 6].

8.7.2.4 Heat balance

Equating heat input to output, the value of I may be calculated from equation (8.5):

$$I = \sqrt{\frac{[387.1(T_m - T_a)(Vd)^{0.448}] + 1.791 \times 10^{-7} \times \varepsilon d[(T_m + 273)^4 - (T_a + 273)] - YSd}{R_{20}[1 + \alpha(T_m - 20)]}} \tag{8.5}$$

A worked example may make these matters become clearer.

Example

Hazel conductor is a 7/3.30 mm AAAC. Assuming that the mechanical considerations such as sag and thermal considerations such as fitting and conductor life are not affected, how much extra current can this conductor take by raising its operating temperature from its normal 50 °C maximum to 75 °C?

The current rating can be calculated for a temperature of 50 °C. For the purpose of this calculation it is assumed that the ambient temperature is 25 °C, there is a cross-wind speed of 0.447 m/s and solar radiation of 580 W/m^2 and that the conductor surface has been blackened by the effects of pollution. This latter point is to enable us to assume a high emissivity factor of 0.9.

From manufacturer's tables the DC resistance of Hazel is found to be $R_{20} = 0.5498\ \Omega/\text{km}$ and the overall diameter (d) to be 9.90 mm.

So, in summary:

conductor operating temperature, $T_m = 50\,°\text{C}$

ambient temperature, $T_a = 25\,°\text{C}$

solar absorption coefficient (blackened surface), $Y = 0.9$

intensity of solar radiation, $S = 580\ \text{W/m}^2$

emissivity of the conductor (blackened surface), $\varepsilon = 0.9$

wind speed, $V = 0.447\ \text{m/s}$

Hazel is an AAAC conductor, so from Table 8.3 we have the temperature coefficient of resistance at 20 °C for aluminium alloy $\alpha = 0.0036$ per °C.

So using equation (8.5):

$$I = \sqrt{}((18842.8 + 4784.8 - 5167.8)/(0.60918))$$

Therefore, the current rating (I) of hazel at an operating temperature of 50 °C is 174 A.

With the operating temperature (T_m) raised to 75 °C the equation becomes:

$$I = \sqrt{}((37685.6 + 10819.4 - 5167.8)/0.65866)$$

Therefore, the current rating (I) of hazel at an operating temperature of 75 °C is 257 A. The temperature increase therefore allows an extra 82 A of current which may be used in a fault situation or for circuit switching.

8.7.3 Continuous current rating for covered conductors

Although we will not deal with covered conductors in detail until section 9.4, it is convenient here to consider the effect that having a sheath will have on the current carrying capacity. When a conductor has a plastic sheath (e.g. PVC or XLPE) the above method of calculating the current rating is not applicable. This is because a temperature gradient will be set up across the covering sheath material. When the conductor itself is operating at its maximum temperature, the temperature at the surface of the plastic will be lower and this will affect the convection, radiation and absorption components of the calculation (Figure 8.5).

There are several methods of calculating the current carrying capacity of conductors – and the correct one is not always the one used. There can be several reasons for this, but in the UK the main reason is historical. There has been a specific way to calculate conductor capacity (known as P27 [7]) for several decades. OHL engineers are used to this method and it is consistent to continue to use it to obtain true comparison with previous conductor values. If a different method is used to calculate the present range of conductors, then confusion can occur if a line is to be refurbished while maintaining the equivalent current carrying capacity. The UK will, however, have to come into line with the European directives on overhead lines and will therefore have to get used to the new techniques. At the risk of adding to the confusion, the method presented in this section is that historically used in the UK. In section 8.7.4 the Cigré-based method will be described.

The following iterative method is not particularly elegant, but it works. Two equations must be satisfied for steady-state conditions. The first is derived from the heat balance equation and is similar to equation (8.5), except that d_o is the overall diameter of the covered conductor, and the convection/radiation terms have been changed to depend on T_s, the surface temperature, as opposed to T_m, the maximum

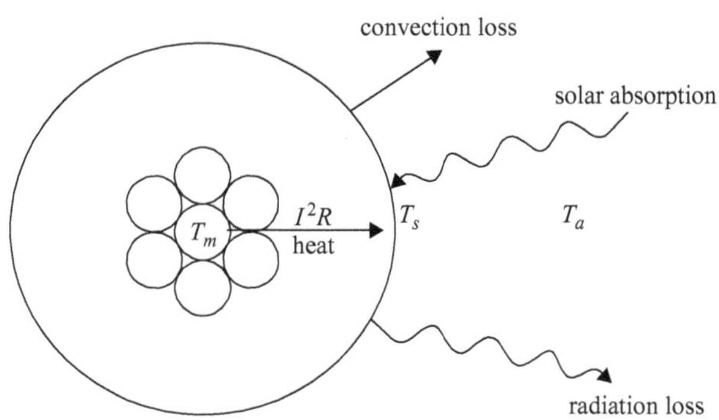

Figure 8.5 Heat energy in/out of a covered conductor

conductor temperature:

$$I = \sqrt{\frac{[387.1(T_m - T_a)(Vd_o)^{0.448}] + 1.791 \times 10^{-7} \times \varepsilon d_o[(T_m + 273)^4 - (T_a + 273)] - YSd_o}{R_{20}[1 + \alpha(T_m - 20)]}} \tag{8.6}$$

The second equation is derived by considering the heat transference through the sheath material:

temperature difference = heat transferred × thermal resistance

$$T_m - T_s = 10^{-3}\{I^2 R_{20}[1 + \alpha(T_m - 20)]\} \times \left[\left(\frac{\rho}{2\pi}\right)\log_e\left(1 + \frac{2b}{d_c}\right)\right] \tag{8.7}$$

where ρ is the thermal resistivity of the sheath (°K m/W) (3.5 for XLPE and 5.5 for PVC), b is the radial thickness of the sheath (mm) and d_c is the conductor diameter (mm).

The steps to be followed to calculate the current rating of the covered conductor are:

1 Use equation (8.6) to calculate I, with T_s set to the maximum allowable conductor temperature, T_m.
2 Use this value of I in equation (8.7) to calculate a new value for T_s.
3 Use this value of T_s in equation (8.6) to calculate a new value for I.
4 Repeat steps (2) and (3) until the values for T_s and I stabilise.

8.7.3.1 Worked example

To do a specific worked example, the current rating of an XLPE sheathed covered conductor can be determined, given that the maximum conductor operating temperature is 75 °C, solar radiation is 850 W/m^2, wind speed is 0.447 m/s and the ambient air temperature is 25 °C:

conductor diameter = 8.6 mm

XLPE radial thickness = 2.5 mm

overall diameter = 13.6 mm

DC resistance at 20 °C = 0.6585 Ω/km

temperature coefficient of resistance = 0.0036/°C

emissivity = 0.95

absorption coefficient = 0.95

1 Setting T_s = 75 °C, equation (8.6) gives:

$$I = 247.1\text{ A}$$

2 Using this value of I in equation (8.7) we obtain:

$$T_s = 62.7\,°\text{C}$$

3 Using this value to re-calculate I from equation (8.6) gives:

$$I = 204.3\ \text{A}$$

4 Repeating the above steps until the values of T_s and I stabilise gives:

$$T_s = 65.7\,^{\circ}\text{C}$$

$$I = 215.3\ \text{A}$$

The current rating of the cable is therefore 215.3 A.

8.7.4 Alternative current carrying capacity calculation method for covered conductors

It has become rather confusing to get a whole picture of current carrying capacities of covered overhead lines. Current ratings of similar conductor constructions vary depending where they are calculated. Especially, there have been some major differences between the calculations made in Scandinavia and UK. The Scandinavian way follows current European thinking and will most likely form the basis for new legislation. The current rating calculation presented here now is based mainly on the method devised by Cigré Working Group 22.12 and published in *Électra* no. 144 [2]. However, as this method was developed for bare overhead conductors, some modifications have been made to take account of the sheath. The author is grateful to Mr Jaako Pitkänen of Pirelli Cables and Systems Oy, Pikala, Finland for providing the following detailed calculation.

It is obvious that different calculation methods will give slightly different current ratings for the same kinds of conductors, but this difference is not always fully due to the differences between the calculation methods. There are also different environmental parameters considered in the calculations.

8.7.4.1 Principles of calculation method used

As in section 8.7.3, the current carrying capacity calculation method is based on a simple heat balance equation

$$\text{heat gain} = \text{heat loss}$$

which is formulated as:

$$P_J + P_S = P_R + P_C \tag{8.8}$$

where P_J is joule losses generated in the conductor, Wm^{-1}, P_S is solar heating, Wm^{-1}, P_R is radiative cooling, Wm^{-1} and P_C is convective cooling, Wm^{-1}.

Joule heating can be calculated by means of the AC resistance of the conductor:

$$P_J = I^2 R_{\text{AC}} \tag{8.9}$$

where I is current, A, and R_{AC} is AC resistance of the conductor, $\Omega\,\text{m}^{-1}$.

Thus the current can be obtained from equations (8.8) and (8.9):

$$I = \sqrt{\frac{P_R + P_C - P_S}{R_{\mathrm{AC}}}} \tag{8.10}$$

The current obtained from equation (8.10) can be counted as the current carrying capacity of a covered conductor if the AC resistance, radiative cooling etc. are calculated at the maximum allowable temperature of the conductor.

However, as has already been shown, the calculation of current carrying capacity is an iterative process, because radiative and convective cooling are functions of the surface temperature of the sheath that is itself dependent on the load current. Since the surface temperature is actually a function of joule heating the cooling expressions might be written as:

$$\begin{aligned} P_R &= P_R(P_J) = P_R(I^2 R_{\mathrm{AC}}) \\ P_C &= P_C(P_J) = P_C(I^2 R_{\mathrm{AC}}) \end{aligned} \tag{8.11}$$

When calculating current carrying capacity the iteration steps are as follows:

1 Estimate an approximate value for current I_0.
2 Calculate cooling powers $P_R(I_0^2 R_{\mathrm{AC}})$ and $P_C(I_0^2 R_{\mathrm{AC}})$.
3 Calculate current I_1 from equation (8.10).
4 Calculate new values of cooling powers $P_R(I_1^2 R_{\mathrm{AC}})$ and $P_C(I_1^2 R_{\mathrm{AC}})$ using current I_1 in equation (8.11).
5 Calculate new value of current from equation (8.10).
6 Repeat steps 2 to 5 n times until the differential $|I_n - I_{n-1}|$ is small enough (for example smaller than 0.1 A).

The maximum current carrying capacity of the covered conductor is then finally obtained. This is exactly the same process as in the UK method.

8.7.4.2 Calculation of AC resistance R_{AC} [8]

AC resistance of a covered conductor can be calculated according to the standard IEC 287-1-1:1994 using sub-clauses 2.1, 2.1.1 and 2.1.2 [8]. However, the proximity effect factor described in sub-clause 2.1 can be neglected in the case of widely spaced covered overhead conductors.

8.7.4.3 Calculation of radiative cooling P_R

From the Cigré method [2] the formula for radiative cooling is:

$$P_R = \pi D \varepsilon \sigma_B[(\theta_S + 273)^4 - (\theta_a + 273)^4] \tag{8.12}$$

where

D = outer diameter of the covered conductor, m

ε = emissivity coefficient

σ_B = Stefan–Boltzman constant, $5.67051 \cdot 10^{-8}\ \mathrm{Wm^{-2}K^{-4}}$

θ_S = surface temperature of sheath, °C

θ_a = ambient temperature, °C

The main problem when using formula (8.12) is that the surface temperature of the conductor is not known. In some cases it may be assumed that the surface temperature of the sheath is a few °C below the conductor temperature. However, an exact approach is given next.

8.7.4.4 Calculation of surface temperature of the sheath

By using the analogy between electrical circuit theory and heat transfer phenomena, the surface temperature can be calculated as follows:

$$\theta_S = \theta_C - T_3 P_J \tag{8.13}$$

where T_3 is the thermal resistance of sheath, $\mathrm{KmW^{-1}}$.

From IEC 60287 [9] we can get a formula for the thermal resistance of the sheath:

$$T_3 = \frac{\rho_T}{2\pi} \ln\left(1 + \frac{2t_1}{d_c}\right) \tag{8.14}$$

where ρ_T is the thermal resistivity of sheath, $\mathrm{KmW^{-1}}$, t_1 is the thickness of sheath, mm and d_c is the diameter of AlMgSi conductor, mm.

8.7.4.5 Radiative cooling as a function of joule heating

As mentioned earlier, the radiative cooling is a function of the joule heating, P_J, and it can be obtained by substituting equation (8.13) into equation (8.12):

$$P_R(P_J) = \pi D \varepsilon \sigma_B[((\theta_C - P_J T_3) + 273)^4 - (\theta_a + 273)^4] \tag{8.15}$$

8.7.4.6 Calculation of convective cooling P_C

From the Cigré calculation method [2] we get a formula for convective cooling

$$P_C = \pi \lambda_f (\theta_S - \theta_a) Nu \tag{8.16}$$

where λ_f is the thermal conductivity of air, $\mathrm{Wm^{-1}K^{-1}}$, θ_S is the surface temperature of the sheath, °C, θ_a is the ambient temperature, °C and *Nu* is the Nusselt number.

An empirical value for the thermal conductivity of air is [2]:

$$\lambda_f = 2.42 \cdot 10^{-2} + 7.2 \cdot 10^{-5} \cdot 0.5(\theta_S - \theta_a) \tag{8.17}$$

Convective cooling is often divided into natural and forced cooling. It can safely be assumed that cooling is forced by the wind in this case. It is also more convenient to assume that the wind direction is normal to the span (angle of wind attack is 90°) and that the wind speed is more than 0.5 m/s.

The Nusselt number for forced cooling that is valid in these circumstances can be calculated from equation (8.18). If it is necessary to calculate the Nusselt number in other conditions than those mentioned earlier the relevant formulae can be found [2]:

$$Nu = B_1 (Re)^n \tag{8.18}$$

where Re is the Reynolds number and B_1 and n are coefficients depending on the Reynolds number and roughness of the sheath's outer surface.

The Reynolds number can be calculated from equation (8.19) [2]:

$$Re = \frac{\rho_r v D}{\nu} \tag{8.19}$$

where ρ_r is the relative air density, v is the wind velocity, $\mathrm{ms^{-1}}$, D is the outer diameter of the covered conductor, m and ν is the kinematic viscosity of air, $\mathrm{m^2s^{-1}}$.

Empirical equations for relative density of air and for kinematic viscosity of air are [2]:

$$\rho_r = e^{(-1.16 \cdot 10^{-4} y)} \tag{8.20}$$

$$\nu = 1.32 \cdot 10^{-5} + 9.5 \cdot 10^{-8} \cdot 0.5(\theta_S + \theta_a) \tag{8.21}$$

where y is the height above sea level (m).

Values for coefficients in equation (8.18) are shown in Table 8.7.

8.7.4.7 Convective cooling as a function of joule heating

Convective cooling is a function of the surface temperature of the sheath in the same way as radiative cooling. This can be seen especially from equations (8.9), (8.10) and (8.14). Because the surface temperature is a function of joule heating, the convective cooling might be written in terms of joule heating.

8.7.4.8 Calculation of solar heating P_S [2]

The formula for solar heating using global solar radiation can be obtained from equation (8.22):

$$P_S = \alpha_s S D \tag{8.22}$$

where S is the global solar radiation, $\mathrm{Wm^{-2}}$, D is the outer diameter of covered conductor, m and α_s is the absorptivity of sheath surface.

Table 8.7 Constants B_1 and n as a function of Reynolds number [2]

Reynolds number		B_1	n
from	to		
10^2	$2.65 \cdot 10^3$	0.641	0.471
$2.65 \cdot 10^3$	$5 \cdot 10^4$	0.178	0.633

Values given in the table are valid for covered conductors with surface roughness of less than 0.05. That is why the given values are suitable for the UK smooth-surfaced covered conductors.

8.7.5 Choosing the parameters for calculation

In previous sections different parameter values have been used. The first parameter, which may differ in calculations made by different users of covered conductors, is the emissivity. In Reference 2 the suggested value is $\varepsilon = 0.95$ for weathered bare overhead lines. From earlier in this section, the value of the emissivity is taken as around 0.9 for black surfaces. In essence, the difference has very little effect. The big difference is only between a new aluminium conductor and an aged or covered conductor. Thus it is reasonable to use a value of 0.95 for black surfaced covered conductors. If a green or grey sheath is used then a lower emissivity value should be used.

The second parameter is the absorptivity of the normal sheath material (XLPE) of covered conductors. It is suggested here to be close to the value given for PE in standard IEC 287-2-1:1994 Table 7.2 [9]. The value of absorptivity for PE in that standard is 0.4. Nevertheless, it is stated in Reference 2 that the absorptivity for weathered bare overhead line conductor in an industrial environment is as high as 0.95. There is some discrepancy between these approaches. Thus, the conservative but acceptable value for absorptivity might be 0.5. There are also different surface materials such as the Swedish HDPE that is coloured green.

Other parameters such as wind velocity, intensity of solar radiation, highest operating temperature of the conductor etc. vary from country to country. The values standardised in Scandinavia are as follows [10]:

solar intensity: 1000 Wm^{-2}
wind velocity: 0.6 ms^{-1}
attack angle of wind: +90°
air temperature: +20 °C
maximum permitted operation temperature: +80 °C.

The values of other parameters used in current rating calculations in this section are

absorptivity: 0.5
altitude from sea level: 50.0 m
emissivity factor: 0.95
thermal resistance of PE: 3.5.

8.7.5.1 Current ratings for covered conductors

Current ratings for the most commonly used cross-sections of AAAC covered conductors for medium voltage are listed in Table 8.8. Current ratings are calculated based on the environmental conditions described above.

Table 8.8 Current carrying capacity of XLPE covered conductors

Conductor size (mm^2)	240	185	150	120	95	70	50	35
Current carrying capacity (A)	662	561	487	430	371	307	243	200

8.7.6 Short circuit current rating

To determine the maximum current that can be taken safely by a conductor under a fault condition, the duration of the fault and the highest temperature that the conductor may be allowed to reach must both be known.

The limiting factor is that the conductor should not lose a significant amount of its tensile strength due to the annealing effect of the heat generated. The fault duration, normally less than 5 s, is insignificant compared to the cooling time of the conductor, and therefore a single maximum temperature can be set for each conductor material (usually 200 °C for aluminium and 210 °C for copper). However, these may be lower depending on whether the conductor is bare or covered with PVC, PE, XLPE or HDPE. The author is grateful to John Evans for his contribution to these calculations.

Making the following simplifying assumptions:

1 no heat is lost from the conductor
2 the specific heat of the conductor is constant
3 the temperature coefficient of resistance is constant
4 the fault current is constant.

We can equate the electrical energy generated during the fault to the heat energy gained, as follows:

$$\int_0^t I^2 R_1 \times 10^6 (1 + r\eta) dt = \int_0^{T_2 - T_1} WS \times 10^3 dt \tag{8.23}$$

where:

I = short circuit current (kA)

T = duration of short circuit (s)

R_1 = resistance of conductor at initial temperature T_1 (Ω/km)

r = temperature coefficient of resistance (per °C)

η = conductor temperature rise above T_1 (°C)

T_1 = temperature of conductor before short circuit (°C)

T_2 = temperature of conductor after short circuit (°C)

W = mass of conducting material in conductor (kg/km)

S = heat capacity of conducting material (J/°C g)

Equation (8.23) can be re-arranged to give:

$$I^2 t = \left(\frac{WS \times 10^3}{R_1 r} \right) \log_e [1 + r(T_2 - T_1)] \tag{8.24}$$

This is the usual way to express the short circuit rating of a conductor, i.e. its I^2t rating.

Table 8.9 Some useful constants for evaluation of I^2t

Material	r (/°C)	S (J/°C.g)
Aluminium	0.00403	0.932
Aluminium alloy	0.0036	0.932
Steel	0.00537	0.488
Aluminium clad steel	0.005115	0.5215

For conductors constructed with two or more metals, the constants of this equation must be averaged as follows:

$$W = W_1 + W_2 + W_3 + W_n$$

$$S = \frac{W_1 S_1 + W_2 S_2 + W_3 S_3 + W_n S_n}{W}$$

$$r = \frac{A_1 r_1 + A_2 r_2 + A_3 r_3 + A_n r_n}{A}$$

$$R = \frac{1}{1/R_1 + 1/R_2 + 1/R_3 + 1/R_n}$$

where A is the cross-sectional area (mm^2).

Some useful constants are given in Table 8.9.

Going back to Hazel conductor again, the short circuit (I^2t) rating of this conductor operating at 50 °C can now be calculated, given a maximum tolerable temperature of 200 °C.

In the calculations the following relevant properties of Hazel are assumed:

- ungreased mass of 164 kg/km
- heat capacity of 0.932 J/°C.g
- resistance at 20 °C of 0.5498 Ω/km
- temperature coefficient of resistance (at 20 °C) of 0.0036/°C.

The resistance R of the conductor at 50 °C is:

$$R_1 = R_{20}[1 + r(50 - 20)]$$

$$R_1 = 0.5498[1 + (0.0036 \times 30)]$$

$$R_1 = 0.6092\,\Omega/\text{km}$$

Applying equation (8.24) with:

$$W = 164\,\text{kg/km}$$

$$S = 0.932\,\text{J/°C.g}$$

$$R_1 = 0.6092\,\Omega/\text{km}$$

$$r = 0.0036/^\circ\text{C}$$
$$T_2 = 200\,^\circ\text{C}$$
$$T_1 = 50\,^\circ\text{C}$$

gives:

$$I^2t = \frac{164 \times 0.932 \times 10^{-3}}{0.6092 \times 0.0036} \log_e[1 + 0.0036(200 - 50)]$$
$$I^2t = 30.1\,\text{kA}^2\text{s}$$

In one sense, Hazel is an easy example as it is a one-material conductor. A look at an ACSR conductor (Rabbit) of around the same size shows the effect of a conductor composed of two materials.

Rabbit is an ACSR (6/3.35 mm Al + 1/3.35 mm St) conductor. Assume an operating temperature up to 65 °C. Given a protection system response time of 0.5 s and maximum tolerable short-term temperature of 200 °C, it is possible to calculate the maximum permitted fault current that the conductor can withstand.

It is worthwhile to determine the effect of the steel core by first of all calculating the short circuit current for the aluminium and then for the whole conductor.

Considering the aluminium component only, first calculate the I^2t rating for Rabbit using the following relevant properties:

- aluminium mass of 145 kg/km
- aluminium heat capacity of 0.932 J/°C.g
- resistance at 20 °C of 0.5426 Ω/km
- temperature coefficient of resistance (at 20 °C) of 0.00403/°C.

The initial resistance (R_1) of the conductor at 65 °C is:

$$R_1 = R_{20}[1 + r(65 - 20)]$$
$$R_1 = 0.5426[1 + (0.00403 \times 45)]$$
$$R_1 = 0.6410\,\Omega/\text{km}$$

Applying equation (8.24) with:

$$W = 145\,\text{kg/km}$$
$$S = 0.932\,\text{J}/^\circ\text{C.g}$$
$$R_1 = 0.6410\,\Omega/\text{km}$$
$$r = 0.00403/^\circ\text{C}$$
$$T_2 = 200\,^\circ\text{C}$$
$$T_1 = 65\,^\circ\text{C}$$

gives:

$$I^2t = \frac{145 \times 0.932 \times 10^{-3}}{0.641 \times 0.00403} \log_e[1 + 0.00403(200 - 65)]$$

$$I^2t = 22.7\,\mathrm{kA^2s}$$

For a fault duration of 0.5 s, this gives a maximum permitted fault current:

$$I = \surd(22.7/0.5)$$

$$I = 6.7\,\mathrm{kA}$$

So far this is a repeat of the Hazel calculation. Now for the two component materials, considering aluminium and steel components.

Again, first calculate the I^2t rating using the following relevant properties of Rabbit:

- total ungreased mass of 214 kg/km (145 kg/km aluminium and 69 kg/km steel)
- total cross-sectional area of 61.69 mm^2 (52.88 mm^2 aluminium and 8.81 mm^2 steel)
- heat capacity of

$$\frac{145 \times 0.932 + 69 \times 0.488}{214} = 0.789\,\mathrm{J/^\circ C.g}$$

- resistance at 20 °C of

$$1/\{(1/0.5426) + (1/(192/8.81))\} = 0.5294\,\Omega/\mathrm{km}$$

- temperature coefficient of resistance (at 20 °C) of

$$\frac{(52.88 \times 0.00403) + (8.81 \times 0.00537)}{61.69} = 0.00422 \text{ per } ^\circ\mathrm{C}$$

The initial resistance (R_1) of the conductor at 65 °C is:

$$R_1 = R_{20}[1 + r(65 - 20)]$$

$$R_1 = 0.5294[1 + (0.00422 \times 45)]$$

$$R_1 = 0.6299\,\Omega/\mathrm{km}$$

Applying equation (8.24) with:

$$W = 214\,\mathrm{kg/km}$$

$$S = 0.789\,\mathrm{J/^\circ C.g}$$

$$R_1 = 0.6299\,\Omega/\mathrm{km}$$

$$r = 0.00422/^\circ\mathrm{C}$$

$$T_2 = 200\,^\circ\mathrm{C}$$

$$T_1 = 65\,^\circ\mathrm{C}$$

gives:

$$I^2t = (214 \times 0.789 \times 10^{-3})/(0.6299 \times 0.00422) \log_e[1 + 0.00422(200 - 65)]$$

$$I^2t = 28.6\,\text{kA}^2\text{s}$$

For a fault duration of 0.5 s, this gives a maximum permitted fault current:

$$I = \frac{\sqrt{28.6}}{0.5}$$

$$I = 7.6\,\text{kA}$$

So the answer is a little different when both components are taken into account. The steel thus has a significant electrical role to play as well as its main strength role.

8.8 Reactance

8.8.1 General

Understanding the behaviour of any system that includes variable currents in a conductor, either as an alternating current or as transients, involves some knowledge of the reactance of the circuits. The intention here is to give a brief introduction to methods of calculating the inductive and capacitative reactances related to overhead conductors in simple configurations. This is not meant to be an exhaustive treatment.

8.8.2 Inductive reactance

The calculation of reactance is simplified if it is to be split into two sources:

1 that due to magnetic flux within a radius of 0.3 m, including the internal reactance within the conductor (X_a)
2 that due to the flux between 0.3 m radius and the centre of the equivalent return conductor (X_d).

The sum of the two terms ($X_a + X_d$) is the inductive reactance of the conductor under normal load conditions. The values X_a and X_d are also used in calculating the zero sequence impedance (under fault conditions).

The value of X_a is an inherent electrical property of the conductor (allowing for the fact that the flux out to 0.3 m is included). The value of X_d, however, depends on the spacing of the conductors and is unrelated to the conductor size. This is an important difference.

8.8.2.1 Calculation of X_a

This calculation is aided by using the factor geometric mean radius (GMR), which represents the radius of an infinitely thin tube, the inductance of which equals that of the conductor.

For a frequency of 50 Hz:

$$X_a = 0.1447 \log_{10}(0.3/\text{GMR}) \tag{8.25}$$

Table 8.10 Values of GMR for some common conductor types (d = conductor overall diameter (m))

Aluminium		Steel		GMR (m)
wires	layers	wires	layers	
7	1	–	–	$0.3628d$
19	2	–	–	$0.3789d$
37	3	–	–	$0.3839d$
61	4	–	–	$0.3861d$
6	1	1	–	$0.2545d$[1]
12	1	7	1	$0.165d$[1]
30	2	7	1	$0.4072d$
54	3	7	1	$0.4050d$

[1] These values are approximate only and depend on the current load.

where X_a is the inductive reactance (Ω/km) and GMR is the geometric mean radius (m).

The GMR depends on the diameter and stranding construction of the conductor. For common conductor types, the GMR may be calculated from Table 8.10.

8.8.3 Capacitive reactance

Capacitive reactance can also be divided into two components:

1 the capacitance out to a radius of 0.3 m (X'_a)
2 the capacitance from 0.3 m to the equivalent return conductor (X'_d), and depends on the geometry of the line

$$X'_a = 0.1319 \log_{10}(0.6/d) \tag{8.26}$$

where X'_a is the capacitative reactance (MΩ km) and d is the conductor diameter (m).

As an example, to calculate the inductive and capacitative reactance to 0.3 m (X_a) for UPAS (37/3.53 mm AAAC), we use Table 8.10 and equations (8.25) and (8.26). The diameter of UPAS is 24.71 mm.

From Table 8.10:

$$\text{GMR} = 0.3839 \times 0.02471 = 0.009486\,\text{m}$$

So the inductive reactance is given by:

$$X_a = 0.1447 \log_{10}(0.3/0.009486) = 0.217\,\Omega/\text{km}$$

and the capacitive reactance to a radius of 0.3 m from UPAS AAAC is given by:

$$X'_a = 0.1319 \log_{10}(0.6/0.02471) = 0.183\,\text{M}\Omega\,\text{km}$$

8.9 References

1 Cigré study committee 22-W904: 'Recommendations for the evaluation of the lifetime of transmission line conductors', *Electra*, 1979 (63), pp. 103–145
2 Cigré WG 22.12: 'The thermal behaviour of overhead conductors, section 1 and 2: mathematical model for evaluation of conductor temperature in the steady state and the application thereof', *Electra*, 1992 (144), pp. 43–45
3 Cigré WG 22.12: 'The thermal behaviour of overhead conductors section 4: mathematical model for evaluation of conductor temperature in the adiabatic state', *Electra*, 1999 (185), pp. 75–88
4 ENA Engineering Recommendation L38/1
5 MORGAN, V. T.: 'Electrical characteristics of steel-core aluminium conductors', *IEE Proc.*, 1965, **112**(2)
6 The Aluminium Association: 'Aluminium electrical conductor handbook', 1988
7 ENA Engineering Recommendation P27: 'Current rating guide for high voltage overhead lines operating in the UK distribution system'
8 IEC Standard: 'Electric cables – calculation of the current rating – part 1: current rating equations (100% load factor) and calculation of losses – section 1: general'. IEC Publication 60287-1-1, 1994
9 IEC Standard: 'Electric cables – calculation of the current rating – part 2: thermal resistance – section 1: calculation of thermal resistance'. IEC Publication 60287-2-1, 1994
10 Finnish Electrotechnical Standards Association: '12/20 kV overhead lines. PAS-system'. Standard SFS 5790, 1995

Chapter 9

Bare, insulated and covered conductors

9.1 Introduction

This chapter looks briefly at the various types of conductor on the market with particular reference to covered and insulated conductors. Aspects such as current carrying capacity, weather loads and line design have been covered in other chapters. A brief introduction to novel conductors that may be used in the future on wood pole lines is given at the end of the chapter.

9.2 Types of overhead distribution system

There are four basic types of overhead line system in use at distribution voltages:

1 bare wire
2 XLPE/HDPE covered conductors
3 aerial cable systems
4 spacer cable concept.

XLPE (cross-linked polyethylene) and HDPE (high density polyethylene) are the two materials of which the sheaths of covered conductors are made.

The range of characteristics that are normally of most importance are:

- electrical
- mechanical
- size
- chemical/physical
- cost
- reliability
- environmental.

'Environmental' covers susceptibility to wind/ice/snow etc. as well as pollution such as salt.

9.2.1 Electrical

- conductor type – high or low conductivity
- current ratings
- thermal uprating.

The conductor material can be a high conductivity copper or aluminium or a conductor designed to give a balance between strength and conductivity. The terms used to describe conductors are given in chapter 8, as well as the thermal treatments that can help to give the desired characteristics. Thermal uprating is of importance as an alternative to re-conductoring or raising the voltage level when higher power levels are required. If a conductor can be raised in temperature without a reduction in lifetime or an increase in sag so much that it contravenes regulatory clearances, then this can be a costeffective way to increase power flow.

9.2.2 Mechanical

Mechanical characteristics are required in order to undertake sag/tension calculations and pole loadings. A typical data sheet for 120 mm^2 covered conductor is shown in Table 9.1. The use of these characteristics in line design is covered in chapter 5.

9.2.3 Size

The size of a conductor depends essentially on the two basic requirements:

1 In full span applications the conductor must have sufficient mechanical strength to support its own weight plus the loads due to regional weather conditions (ice and wind).

Table 9.1 Mechanical details of 120 mm² covered conductor

Conductor code name (if any)	120CC35
Covered conductor weight (kg/m)	0.43
Cross-sectional area of conductor (mm^2)	120
Conductor overall diameter (mm)	18.1
Coefficient of linear expansion (/°C)	0.000023
Modulus of elasticity (kg/mm^2)	6150
Rated breaking strength of conductor (kgf)	3609.8
Basic/recommended span (m)	120
Wind pressure on conductor (N/m^2)	380
Radial ice thickness (mm)	9.5
Absolute maximum working tension (MWT) limit (kgf)	1804.9
Temperature at MWT limit (°C)	−5.6
Maximum everyday tension (EDT) limit (kgf)	722.0
Temperature at EDT limit (°C)	5

2 The conductor must be able to carry normal current levels all the time and fault current levels for a time dependent on the circuit parameters.

9.2.4 Chemical/physical

It is not intended to go into the chemical and physical characteristics of conductors in this chapter as they are considered in detail in chapters 7 and 8. Basically, the most common types covered in these chapters are:

- ACSR
- AAC
- AAAC
- ACAR
- HD copper
- cadmium copper.

There are many more types available – especially at transmission voltages where thermal rating is normally of more importance.

9.2.5 Cost and reliability

Figure 9.1 gives a rough indication of the cost and reliability of commonly available bare and covered conductor systems. Bare wire is normally the cheapest and easiest system to build. Covered conductors have a higher reliability level as clashing and accidental contact is not a problem. Spacer cable is essentially a covered conductor held in a cradle and supported by a steel or alumoweld cable (see later in this chapter). The fact that the conductors are not self-supporting means that vibration risks are reduced and reliability increased. Aerial cable is essentially a fully insulated three-core cable used for overhead applications. Its low susceptibility to lightning and the inherent reliability of the cable design means that this system is the most expensive, but also the most reliable.

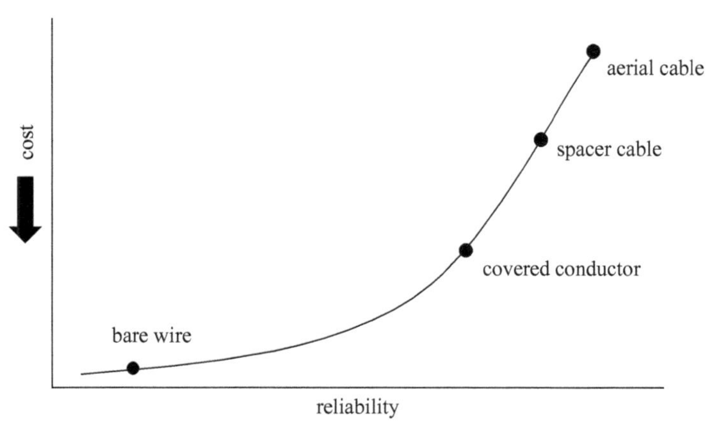

Figure 9.1 Cost and reliability of various conductor systems

Figure 9.2 Line failure when the conductor can withstand the load but the pole cannot

9.2.6 Environmental

Several characteristics affect the environmental performance of a conductor:

- size
- strength
- interstices
- ice/snow
- wind
- corrosion.

The size of a conductor has an influence on the wind effect and the ice and snow loads likely to accumulate, and its strength determines its mechanical withstand capability. Corrosion, especially due to salt, can start by the creation of electrolytic cells in the interstices between the strands where oxygen levels are low. The effect of the environment on overhead lines is well known and if the conductor is too strong then something else has to give (Figure 9.2).

This chapter looks mainly at the mechanical characteristics, the reliability and the effect of salt pollution.

9.3 Bare wire

Bare wire systems have the following characteristics:

- lowest initial material cost
- depend entirely on physical separation of conductors and the insulating value of air for resistance to flashovers (Figure 9.3)
- provide no resistance to outages caused by clashing or contact with trees or wildlife
- utilise cross-arms and insulators.

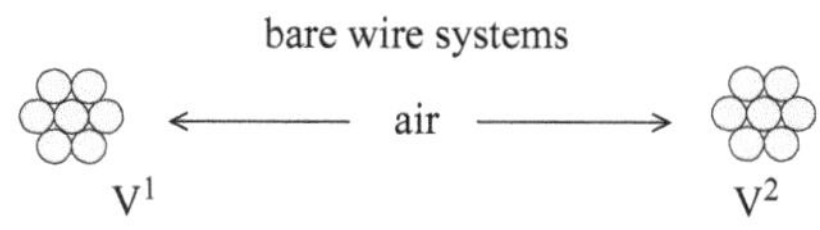

Figure 9.3 Air is the only insulation used for bare conductors in the span

The last point may seem superfluous, but as will be seen later, cross-arms and insulators are not always used.

9.4 Covered conductors

9.4.1 General

This section is basically informative about a brief history of the development of covered conductors. The aim is to provide detailed information on the advantages and difficulties of using covered conductors as an alternative to bare wire overhead lines in the UK. The term 'covered' implies a sheath covering only and not insulated conductors in medium voltage systems (10 to 132 kV). Historically, these conductors have several names such as PAS, BLX, CC, CCT, BLX-T and SAX, but are all basically aluminium alloy conductors with a cross-linked polyethylene (XLPE) coating of 2 to 6 mm. The conductor is also produced, however, with 1.6 and 1.8 mm thick sheaths for ACSR, AACSR and copper conductor. To improve long-term phase-to-phase contact performance at 33 kV, sheath thicknesses of up to 3.3 mm can be used.

9.4.2 Types available

There are several types of covered conductors available in the UK.

9.4.2.1 SAX

The SAX cable is of Finnish manufacturer and was the earliest covered conductor to be widely employed at 20 kV in Scandinavia. Substantial lengths (over 6000 km) have been erected and Finland is still the major user of this type of conductor in Scandinavia, where it comprises eight per cent of the medium voltage network. It is available in 35 to 240 mm^2 sizes. The aluminium alloy conductor is fully compacted and is available ungreased or with powder, tape and grease water blocking, covered with a 2.3 or 3.3 mm thickness of XLPE normally with 2.5 to 3 per cent carbon content.

9.4.2.2 PAS/BLX

PAS/BLX conductors are essentially the same as the SAX system but are manufactured in Norway. The aluminium alloy conductor is normally supplied in compacted form from 35 to 241 mm^2. BLX is also available with copper conductor for use in highly saline areas. It is used on the southern Baltic Norwegian coast. The XLPE is available in nominal thicknesses of 2 to 3 mm.

9.4.2.3 BLX and BLX-T

Another type of covered conductor is known as BLX and manufactured in Sweden. This is a triple extrusion system and is detailed later in this chapter. It is available in low (0.5 per cent) or zero carbon content sheath material to reduce tracking problems. Generally uncompacted, it has a larger overall diameter than the equivalent compacted versions. A mastic compound provides moisture penetration resistance.

9.4.2.4 CC/CCT

Covered conductor (CC) and covered conductor with increased insulation thickness (CCT) are used extensively in Australia and the Far East. They can have a grey XLPE coating compared with the black XLPE of BLX. The lighter CC (2 mm insulation) has tended to be used for rural, long span situations, and the CCT (2.3 mm insulation) is used for forested terrain. Apart from the UV stabilisation (carbon black in the BLX, titanium oxide in the CC and CCT) the other major difference from the PAS/BLX system is that the conductor is uncompacted and is not greased. It also has a water-blocking compound between it and the XLPE layer (ethylene vinyl acetate – EVA). A similar product is made in Sweden.

For simplicity, the SAX, PAS, CC, CCT and BLX conductors will all be referred to as CC in this chapter.

9.4.3 History of covered conductors

Covered conductors have been used in America and Australia for over 40 years. However, the initial coverings of PVC, HDPE and nylon gave very limited lifetimes, suffered surface degradation and were also subject to failure due to lightning damage. Problems with the use of bare overhead lines, especially when conductor breakages did not activate line protection devices, led to increased research into covered lines. Low-voltage ABC came into widespread use, but the expense and other problems associated with the semi-conducting external layer restricted the use of the high-voltage version of this conductor. Safety considerations (both human and wildlife), conductor clashing and tree problems and the generally ageing medium voltage distribution networks led to a re-consideration of the use of covered conductors in the period 1985–90. This process led to their development as one alternative to undergrounding the medium voltage system.

A series of severe bush fires in Australia was blamed on arcing from the clashing of bare OHL conductors. Undergrounding was not considered a generally viable alternative due to frequent disturbances from human activities and tree growth and the high cost of repair. The high costs and short span lengths associated with fully insulated overhead systems such as HV ABC led to the development of CC conductor. The initial use of CC showed up problems of corrosion at mid span and at terminal clamps.

Early water blocking systems using a soft mastic compound caused problems in conductor stripping and high-temperature stability. This was replaced by the Japanese technique of applying a non-conductive layer of ethylene vinyl acetate (EVA) to each layer of the conductor under high temperature and pressure with a final external layer, which is bonded to an XLPE external sheath on extrusion and curing. The completed conductor is intended to provide a watertight seal even at an open bare end without the need for an insulating wrap.

The XLPE thickness of CC is 2.0 mm. This gave a failure rate of 80 days when exposed to abrasion by trees. To extend this period, another version, CCT, was introduced with thicker (2.3 mm) insulation. The increased diameter, however, increases the wind loading and reduces the maximum span length.

In 1987, cable failures in Italy of grey peroxide cross-linked insulated cables were put down to excessive weathering degradation. This led to a more stringent UV stability test standard using radiation down to 235 nm. Polyethylene has a maximum sensitivity to radiation of around 300 nm.

Conventional methods of improving weathering include the addition of carbon black, inorganic pigments and organic stabilisers. However, the Australian requirement of a grey insulation restricted the choices available. A balance had to be struck between good weathering, mechanical strength and expense. In the final choice (named BPH 526) the titanium oxide level was increased by a factor of three compared with HFDM 4292. This satisfied the new Italian standard and, after further trials, was also accepted by the Electricity Trust of South Australia (ETSA).

The large area of forested terrain in Finland led to an early interest in the use of covered conductors, although this was initially concentrated on LV ABC. The interest in CC in Australia led to Finnish research in the late 1970s into covered conductors with the main impetus being the reduction of forest fires caused by trees falling onto bare overhead lines. The result of this work was the introduction of BLX or SAX into the Finnish 20 kV network.

By 1989, Norway and Sweden followed this idea with research on a 22 kV line at Flesberg showing that such conductors could withstand prolonged fallen trees with minimal mechanical problems. There were, however, discharges between the phases and between a phase conductor and the lying trees. At 24 kV this led to an earth fault in one case within three months. The tendency of many UK utilities to have infrequent line inspection means that this sort of incident could lie undetected until damage had occurred to the line. However, this ability to maintain supplies under tree problems would allow maintenance and repair to be carried out more on a planned rather than an emergency basis.

9.4.4 Safety aspects of covered conductor use

9.4.4.1 Saving lives at low voltage

Initially, bare conductors were used on all forms of distribution overhead line construction. At the low-voltage (LV, <1000 V) end, however, much of the overhead line construction erected today is insulated conductor – typically the aerial bundled conductor (ABC) design. This bundled design was also considered for the medium voltage range but was found to be somewhat unreliable due to the high electrical field strength eventually causing breakdown between the covered sheaths of the conductor. These conductors were not screened at the time nor did they have sufficient covering to prevent breakdown between the phases.

9.4.4.2 Saving lives at distribution voltage levels

Perhaps a significant reduction in fatalities and injuries can be achieved in the future by the increased use of CC lines at all voltage levels. Although the use of covered conductors would not have prevented the car-related incidents causing severe injury or death, in several other cases the use of CC would have saved the lives of the people involved. There have been no recorded human fatalities connected with CC lines in the UK since they were first introduced in 1994. Compared to bare wire lines, considering the quantity of CC in use, there would have been seven fatalities expected on these lines within the last six years. However, CC use can be expensive and the safety aspect alone may not be sufficient to justify the additional expenditure.

9.4.4.3 Saving lives at higher voltage levels

The use of low-profile wood pole 132 kV high-voltage covered conductor (HVCC) lines could reduce accidents by:

- being more visible to aircraft
- being lower in height than a tower line
- being safer for a glancing contact or a tangle (e.g. with a hang glider or balloon).

In summary, covered conductors have been in use in Scandinavia for over 20 years and are now being used at voltages up to 132 kV. Together with the UK, over 18 000 route km has been installed at 10–33 kV. This is a large enough sample to see the effects of reducing fatalities as compared with bare wire lines. The results confirm previous experience in Japan – there has been a significant reduction in fatalities to people in both countries, deaths associated with accidental contact being virtually eliminated.

9.4.4.4 Saving wildlife

A CC line is more obvious to flying birds, allowing them more time to deviate from their intended flight path. Birds taking off from lines will also survive as short-term shorting of the phase will have no effect. At present, bare wire lines have warning bells that make the line more obvious to birds and (near airports) to low-flying aircraft. These and other types of bird diverters are not always successful and can be unsightly.

9.4.4.5 Undergrounding

One obvious answer to safety is to underground all overhead lines. But, economics apart, the main reasons for failures in underground systems are that contractors from the gas, telecommunications, water and other utility industries tend to dig them up. It is often very difficult to know which circuit has been dug up as markings may disappear, and network diagrams become uncertain after a few years as landmarks disappear. In chapter 1 the cost of undergrounding was shown to grow considerably as the voltage level rises. Undergrounding only puts electrical conductors out of sight and not necessarily out of contact.

9.4.5 Disadvantages of covered conductors

The major disadvantages of covered conductors are:

Cost

- full installation costs are 10 to 20 per cent more than for bare lines.

Reliability problems

- damage to the line by lightning strike or tree rubbing is difficult to detect
- insulation damage can lead to corrosion problems
- leakage currents at pin insulators can cause insulation damage in polluted environments (e.g. salt fogs)
- Aeolian vibration is greater with BLX/CC/T than with bare lines and could cause premature failure
- care must be taken in the installation and maintenance of lines not to damage the insulation
- discharges between the helically preformed tie and the arc protection device wire and poor connection of the device can both cause radio interference.

Supply quality

- special attention needs to be paid to lightning protection which would not be needed in the case of bare lines
- the frequent use of arc gaps in the lightning protection of BLX/CC/T lines can lead to problems with the quality of supply.

Safety

- some possibilities of bird and squirrel electrocution at pole-top arc gaps
- possible problems with conductor failure detection.

9.4.6 High-voltage covered conductors (HVCC)

Covered conductors used in medium voltages are covered with a 2.3 mm layer of black weather resistant XLPE. At higher voltages a single layer covering is not sufficient due to the higher electrical stresses. This means that when operating above medium voltage levels the partial discharge phenomena inside the covering has to be considered, especially in situations where fallen trees lie on the line or phase conductors are touching each other. Therefore, a new covered conductor type was developed which has a thin layer of extruded semi-conducting XLPE compound controlling the electric field at conductor surface.

The main part of the HVCC covering consists of extruded water–tree retardant XLPE insulation compound. Unfortunately its capability to sustain ultra-violet light, usually from solar radiation, is not very good, which is why there has to be an extra layer applied that guards the main covering against UV light. This outermost layer consists of weather and track resistant black XLPE insulation compound.

All three layers, conductor screen, main covering and outer weather resistant coating, are extruded simultaneously during a highly controlled completely dry curing and cooling process. Thus the triple extruded XLPE covering forms one-piece completeness with various functions.

HVCC SAX consists of watertight, stranded and compacted all aluminium alloy conductor covered by a triple extruded XLPE layer.

A schematic picture of HVCC SAX for voltages from 66 to 132 kV is shown in Figure 9.4.

Covered conductors can have one, two or three sheath layers at medium voltage – at 132 kV the conductor may have five layers. The presence of the sheath (a thin layer of XLPE or HDPE covering – typically 2.3–3.3 mm) allows:

- improved contact outage resistance compared with bare wire
- no clashing problems
- reduced phase spacing
- reduced wildlife problems.

Figures 9.5 and 9.6 show that the presence of the sheath allows the very much closer phase spacing and compact construction typical of CC lines.

As the sheath is insulating (but not insulated) there will be a low-level charging current flowing along its surface. This arises because the sheath forms an insulating

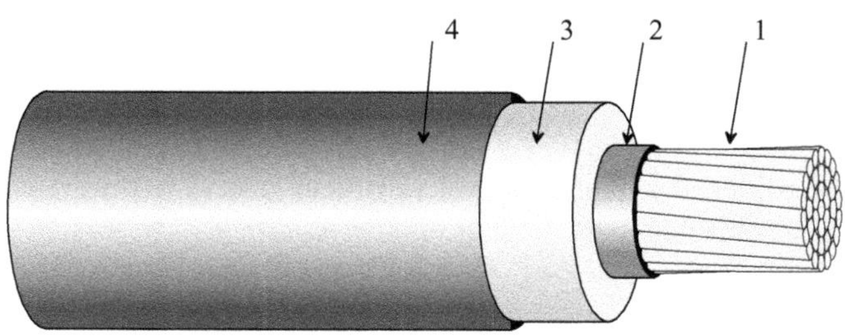

Figure 9.4 *HVCC SAX conductor made in Finland*

1 watertight, stranded and compacted aluminium alloy conductor, cross-sectional areas available from 120 to 355 mm^2

2 extruded semi-conducting XLPE screen

3 extruded water–tree retardant XLPE insulation compound, thickness 2.5…5.5 mm

4 extruded layer of weather and track resistant black XLPE insulation compound, thickness 1 mm

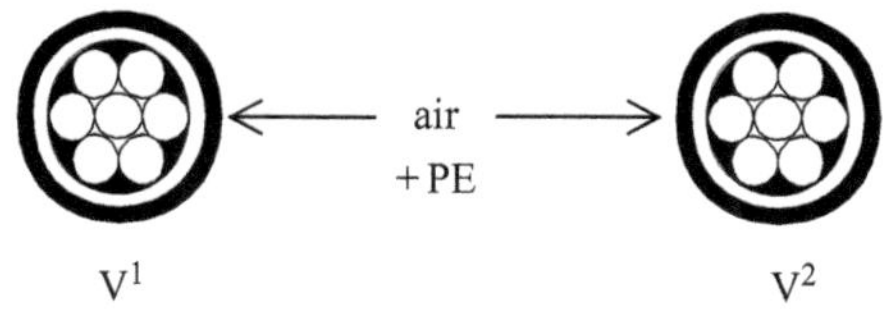

Figure 9.5 The CC sheath allows closer phase spacing

Figure 9.6 The typical close spacing and compact construction of Norwegian CC lines which prompted the first use of CC in the UK

layer between the high-voltage conductor (metal) and the pin or post insulator to earth. This current will normally be less than 0.3 mA. Its characteristics are:

- current inevitably flows phase–phase or phase–ground
- must be low to reduce tracking and erosion, especially under polluted conditions
- metal helical ties form an intermediate electrode and can cause discharge problems (corona) at the ends (if bare).

In cases of surface damage or local pollution, this current can increase sufficiently to cause surface tracking and eventual sheath and subsequent conductor failure. The most common CC in the UK is the single sheath version from Finland shown in Figure 9.7.

The characteristics are:

- single layer
- typically low-density polyethylene
- covering thickness ranges from 2.3 to 3.3 mm

Figure 9.7 Single sheath compacted covered conductor

Figure 9.8 Double layer sheathed CC

Figure 9.9 Triple layer sheathed CC

- lower impulse strength than two- and three-layer designs
- provides some resistance to outages caused by tree and wildlife contact.

The impulse strength of a single layer of XLPE sheathed CC is around 115 kV. Also, the electrical stresses caused by trees on the line or conductors on the cross-arm can erode the sheath in periods from months to minutes depending on the system voltage. Surface voltage stresses are increased if porcelain rather than polymeric insulators are used due to the difference in the dielectric constant of the porcelain (three times that of polymeric insulators). The use of floating helical fittings can also cause surface tracking of an XLPE sheathed conductor in coastal environments especially if the carbon content is around three per cent (which is a common practice). This effect can be reduced by the use of polymeric insulators or switching to an HDPE sheathed conductor that contains substantially less carbon black.

Figure 9.8 shows a double layer CC that is used in the USA but not currently in the UK. This type has higher impulse strength than the single layer.

Figure 9.9 shows a common CC used in Sweden and in some areas in the UK. It is a triple layer with:

- semi-conducting layer
- PE layer
- HDPE or XLPE outside layer.

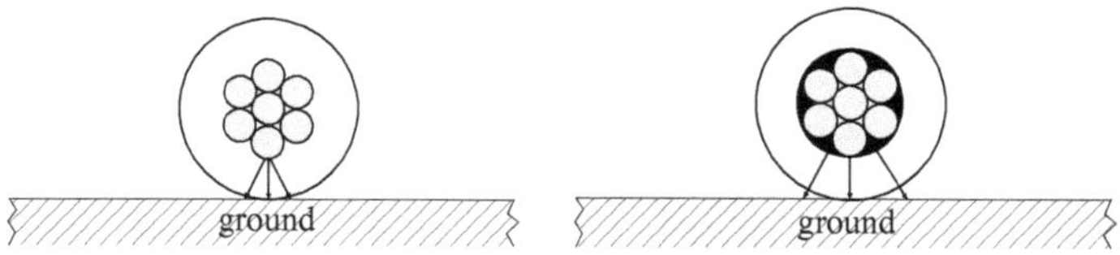

Figure 9.10 *Reduction of voltage stress in an uncompacted conductor by the use of a semi-conductive layer*

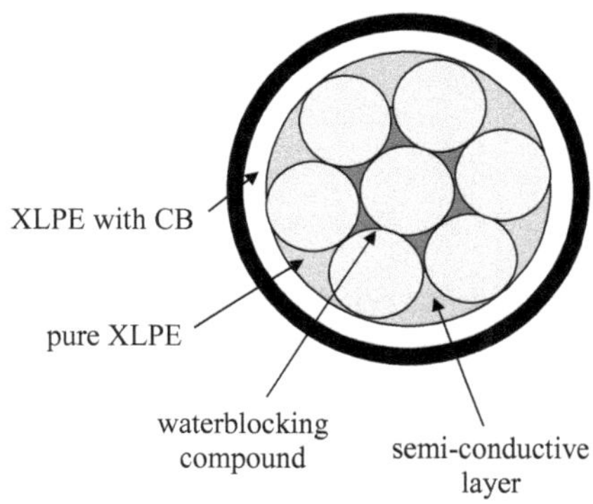

Figure 9.11 *Triple-layer uncompacted CC*

The semi-conducting layer is intended to reduce voltage stresses. The metal strands may be compacted or uncompacted.

The voltage stress is inversely proportional to the strand radius and so a lower voltage stress will occur because of larger effective radius of the whole conductor with the semi-conductive layer as shown in Figure 9.10.

A cross-section of the triple-layer uncompacted CC is shown in Figure 9.11. This conductor had an internal mastic layer to restrict moisture travel along the conductor. Uncompacted conductors have a higher susceptibility to moisture travel than the compacted versions that often use a powder or water-swellable tape to restrict moisture travel. One version of this conductor uses a totally re-cyclable green HDPE outside layer.

9.5 Spacer cable

Spacer cable systems are essentially three CC phases in a polymeric support cradle supported by a messenger cable. Figure 9.12 shows the support system at a pole and Figure 9.13 shows the cable strung at a test site in the UK. Basically, the system has:

- a messenger supported three-layer cable construction in a close triangular configuration
- the mechanical strength to weather severe storms

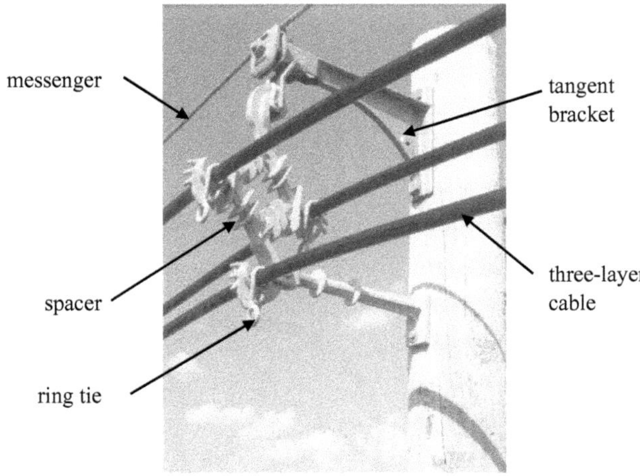

Figure 9.12 Triple sheathed version of the spacer cable (courtesy Hendrix Inc, USA)

Figure 9.13 Spacer cable strung at a UK test site along with other CC and bare wire systems

- the electrical strength to prevent faults due to phase-to-phase or phase-to-ground contact, tree contact or animal contact
- a complete co-ordinated system including cable, messenger, spacers, insulators and hardware.

The conductor may be of the double or triple sheathed CC versions discussed in the previous section. The system is used widely in the USA and parts of Canada and is being marketed in Europe. It has been tested at experimental sites in the UK and some network trials are planned for 2005.

9.6 Aerial cable systems

Aerial cable systems are basically cables that can be strung overhead and run underground and underwater. Such systems obviate the need for OHL/cable junctions and have a very low susceptibility to lightning. They:

- have fully shielded three-core power cables
- have excellent contact resistance
- go overhead, underground, underwater
- have no cross-arms
- have no OHL/underground cable junction
- are made in USA, Sweden and Finland.

Figure 9.14 shows the structure of the Swedish version that is now being used in many parts of the UK. The cable does not use a support wire or messenger, as it is totally self-supporting. It has a high impulse strength of 400 kV and uses an earthed screen. Figure 9.14 shows the aluminium conductor version and Figure 9.15 shows a smaller copper conductor version. The conductor is held up by hooks attached to wood poles in a similar fashion to LV ABC.

Possible uses of aerial cable are:

- replacement of CC overhead lines
- distribution cable in the countryside

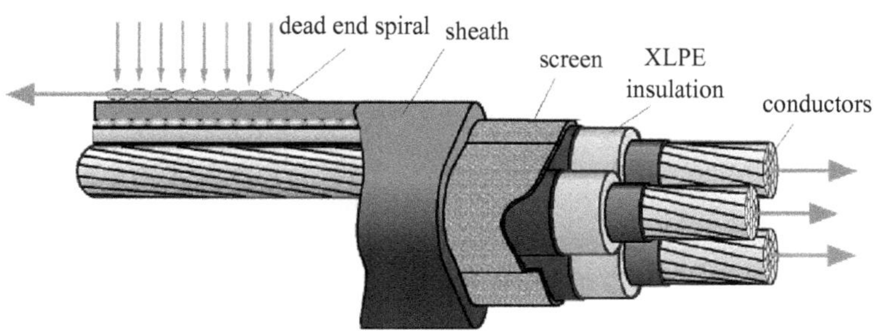

Figure 9.14 The Swedish aerial cable system (courtesy Ericsson Technology Networks Ltd)

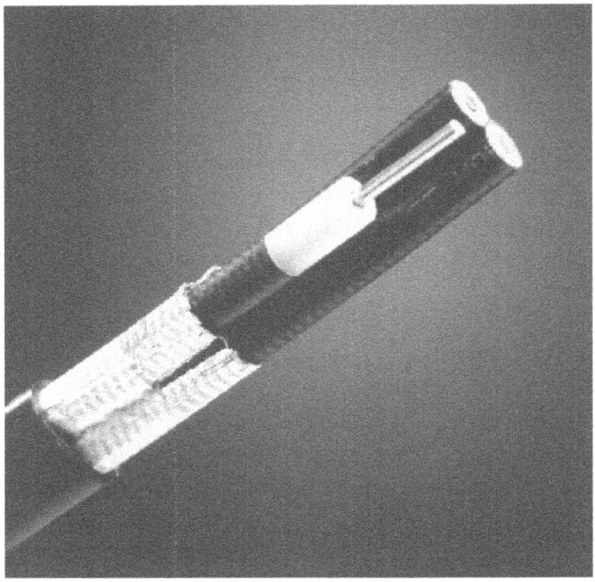

Figure 9.15 The copper conductor version of the aerial cable system

- areas with safety implications with overhead lines near buildings
- temporary installations
- underwater cable
- supply lines that have demand sensitivity
- areas with climatic problems.

Cable systems are more complicated to joint than single-phase OHL conductors and so aerial cable is more suitable for long runs where there are few transformers and spurs lines. It is commonly strung together with LV cable on the same poles.

9.7 PVC conductors

Before 1994 the only covered conductors in widespread use in the UK above 1000 V AC were PVC covered. PVC is one of a family of sheath materials known as thermoplastics. The characteristics of these materials are:

- no chemical bonding of molecules
- many materials available (PVC, polyolefins, polyethylene)
- PE makes up the vast majority of the volume
- a properly designed PE overhead cable will provide long life and excellent outage resistance
- thermoplastic PE cable design is available to provide the same operating temperature rating as XLPE
- lower cost than thermosets.

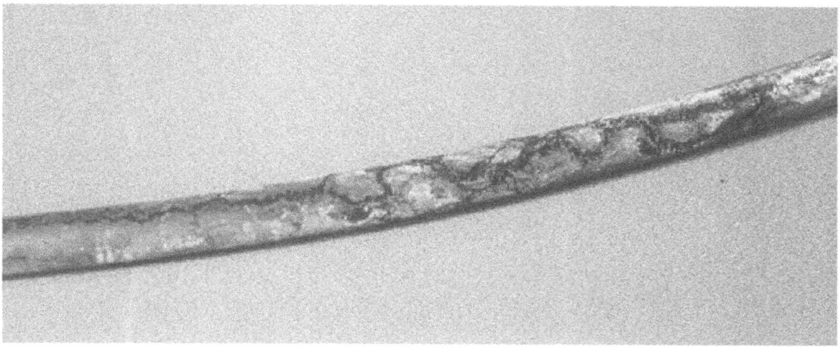

Figure 9.16 Surface tracking on an XLPE covered conductor used with porcelain insulators

The XLPE material is a typical thermoset material. Early PVC covered conductors were susceptible to UV degradation and also responsible for turning water ingress into an acid that then dissolved the metal conductor. PVC sheathed conductors are not normally used now in the UK except for specific local purposes, the developments in thermosets having superseded their use.

9.8 Tracking

Tracking has been mentioned as a problem with covered conductors with high carbon black contents in the sheath material. The carbon black is there as a very effective and cheap UV inhibitor. HDPE materials tend to use the more expensive titanium dioxide as an inhibitor. This does not have the same tendency towards tracking. Tracking results from surface voltage stresses typically between insulation piercing connectors (IPCs) and helical fittings that are not at network voltage. In a highly salt polluted environment, such as within 10 km of the UK coastline, tracking can reduce conductor lifetimes to a few years (Figure 9.16). There are several ways to reduce tracking problems:

1 reducing voltage stress (using polymeric instead of porcelain insulators
2 connecting helical ties with any insulating piercing connectors (IPCs))
3 using crimp connectors instead of electrically floating helical ties
4 using reduced or zero carbon content sheath materials.

9.9 Keeping the power on

Figures 9.17 and 9.18 show real examples of how the power has stayed on when sheathed conductors had been used. Figure 9.17 shows aerial cable where a tree has fallen on the line. The power stayed on and no repairs were required after the tree was removed. Figure 9.18 shows a covered conductor line where a snow-laden branch is

Figure 9.17 A storm in Sweden brings a tree down on an aerial cable line (courtesy Ericsson Technology Networks Ltd)

Figure 9.18 A snow-laden branch resting on a double circuit three-phase CC line in Finland (courtesy Pirelli Cables & Systems Oy)

resting on a three-phase line. If this had been a bare wire line there is no doubt that the power would not have stayed on.

9.10 Novel conductors

9.10.1 Definitions

This section introduces some novel concepts in conductor design [1]. Although initially intended only for transmission lines, they are increasingly being investigated for use at 132 kV and could be used in the future on 66 and 132 kV wood pole lines. The abbreviations used in this section are defined below:

ACSS – aluminium conductor coated steel supported
GTACSR – gap-type TAL aluminium alloy conductor, steel reinforced
HS steel – high strength steel core wires for ACSR
Invar steel – a steel core wire made with high nickel content to reduce the thermal elongation coefficient
Knee-point temperature – the conductor temperature above which the aluminium strands of an ACSR conductor have no tension or go into compression
TACIR – TAL aluminium alloy conductor reinforced with an invar steel core
TACSR – TAL aluminium alloy conductor reinforced by a conventional stranded steel core
TAL – an aluminium zirconium alloy that has stable mechanical and electrical properties after continuous operation at temperatures of up to 150 °C
TW conductor – a bare overhead stranded conductor wherein the aluminium strands are trapezoidal in cross-section
ZTAL – an aluminium zirconium alloy that has stable mechanical and electrical properties after continuous operation at temperatures of up to 210 °C
ZTACIR – ZTAL aluminium alloy conductor reinforced by an invar steel core.

9.10.2 Shaped-strand conductors

Overhead line conductors are normally constructed from helically wound strands of circular cross-section. This gives a conductor cross-section containing large inter-strand voids, with ~20 per cent of the total cross-sectional area of the conductor being air. By using strands with a trapezoidal shape, conductors can be constructed with a higher proportion of metal within their cross-section. Compacted conductors can be homogenous with all strands except the centre strand being of trapezoidal shape, or non-homogenous with a round-stranded, steel core surrounded by trapezoidal aluminium strands. However, the strands may also be ‘Z’ shaped and so effectively lock together.

Shaped-strand conductors have a larger aluminium area and thus lower resistance than a normal round strand conductor with the same outside diameter. When re-conductoring an existing line with shaped-strand conductor, the increased weight of the conductor will result in slightly higher support structure loads, but climatic loads due to wind and/or ice will not be increased, as these are a function of diameter.

For wind-only loading conditions, loads may actually be lower, due to their lower drag coefficients at high wind speeds. One example of a shaped-strand conductor that achieves a low drag coefficient is one that has an oval cross-section, the orientation of which varies along its length, giving a spiral–elliptic shape [2].

Shaped-strand conductors have also been shown to possess slightly better vibration characteristics due to the higher surface area of the contacts between strands of adjacent layers, which results in lower inter-strand contact stresses [3, 4].

9.10.3 Motion-resistant conductors

Shaped-strand conductors have also been used to reduce the effects of wind-induced motions. Such conductors include self-damping conductor (SDC), which incorporates small gaps between the successive layers of strands, allowing energy absorption through impact [5, 6]. Another conductor which resists motion is the T2 conductor, consisting of two standard round conductors wrapped about one another with a helix approximately three metres long [7]. This resists motion due to its aerodynamic characteristics, and is widely used in the United States.

9.10.4 High-temperature conductors

The following four types of conductor are capable of operating continuously at temperatures of at least 150 °C and some as high as 250 °C without significant changes in their mechanical and electrical properties. Each conductor type has certain advantages and disadvantages.

9.10.4.1 Materials

TAL and ZTAL aluminium strands have essentially the same conductivity and tensile strength as ordinary electrical conductor-grade aluminium strand but can operate continuously at temperatures up to 150 °C and 210 °C, respectively, without any loss of tensile strength over time. Fully annealed aluminium strands are chemically identical to ordinary hard drawn aluminium but have much reduced tensile strength. They can operate indefinitely at temperatures even higher than 250 °C without any change in mechanical or electrical properties.

9.10.4.2 Conductor construction

TACSR and (Z)TACIR are stranded in the same fashion as ordinary ACSR. Their electrical and mechanical properties are simply the result of their composite aluminium and steel strand properties.

ACSS can be stranded using either round or trapezoidal-shaped aluminium strands. In either design, the conductor depends primarily on the steel core strands for mechanical strength.

The unique installed properties of G(Z)TACSR are the result of both its strand properties and its construction. The innermost layer of (Z)TAL strands is trapezoidal and a small gap to the core is left to allow installation, with tension applied to the steel core only.

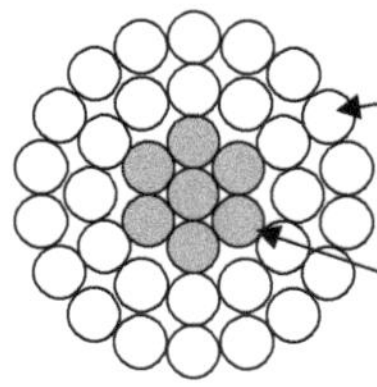

Figure 9.19 Cross-section of TACSR conductor

9.10.4.3 (Z)TACSR

(Z)TACSR has galvanised steel strands for the core surrounded by (Z)TAL strands surrounding them. Figure 9.19 shows the construction.

(Z)TACSR conductor is thus almost identical to conventional ACSR conductors. The main advantage of (Z)TACSR is that its aluminium alloy strands do not anneal at temperatures up to 150 °C for TAL and 210 °C for ZTAL (temperatures above 100 °C would cause annealing of the aluminium strands in a standard ACSR).

(Z)TACSR can therefore be used to uprate existing lines where some additional clearance is available. Steel-cored conductors (and other non-homogeneous conductors) have what is known as a knee-point. This is a temperature above which the higher thermal expansion rate of aluminium causes all the stress of the conductor to be borne by the steel core. Beyond this knee-point temperature, therefore, the conductor experiences a sag increase due to the expansion of steel alone. This new expansion coefficient will be lower than that for the conductor at lower temperatures, resulting in relatively low sag increases when operated at high temperature. Standard ACSR exhibits this property, but usually at a temperature beyond the annealing limit. The TAL alloy of TACSR allows this behaviour to be exploited.

9.10.4.4 G(Z)TACSR

Gap-type conductor [8] has a unique construction. There is small gap between the steel core and the innermost shaped aluminium layer in order to allow the conductor to be tensioned on the steel core only (Figure 9.20). This effectively fixes the conductor's knee-point to the erection temperature, allowing the low-sag properties of the steel core to be exploited over a greater temperature range. The gap is filled with a heat resistant grease to reduce friction between steel core and aluminium layer, and to prevent water penetration.

For a given thermal rating, the G(Z)TACSR will be able to reduce the sag as compared with the conventional ACSR. Special erection techniques are required for gap conductors compared with those of standard construction.

9.10.4.5 (Z)TACIR

As with (Z)TACSR, (Z)TACIR [9] (Figure 9.21) has a conventional ACSR-type stranded construction but makes use of material innovations to give properties allowing the conductor to be operated at high temperatures. In place of the steel strands of (Z)TACSR, it has galvanised or aluminium-clad invar alloy steel wires for the core

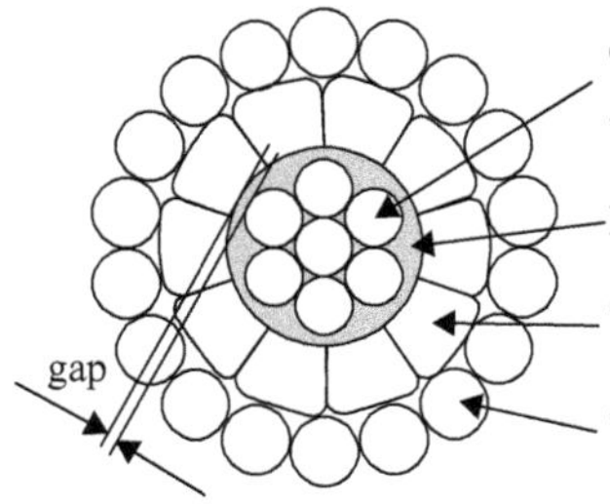

Figure 9.20 Cross-section of GTACSR conductor

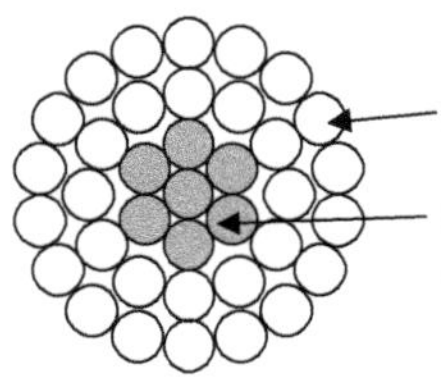

Figure 9.21 Cross-section of (Z)TACIR conductor

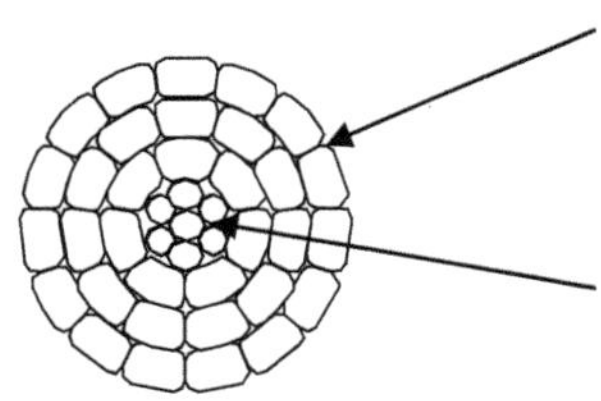

Figure 9.22 Cross-section of ACSS/TW conductor

and (Z)TAL wires surrounding them. Invar is an iron–nickel alloy (Fe/36%Ni) with a very small coefficient of thermal expansion, around one third that of galvanised or aluminium-clad steel wire. The installation methods and accessories for the conductor are virtually the same as those used for conventional ACSR.

9.10.4.6 ACSS and ACSS/TW

ACSS [10] and shaped (trapezoidal) ACSS/TW [11] (Figure 9.22) consist of fully annealed strands of aluminium (1350-0) concentric-lay-stranded about a coated stranded steel core. Special high-strength constructions are also available.

In all designs, the use of annealed aluminium strands yields much higher mechanical self-damping than standard ACSR of the same stranding ratio.

Because the tensile strength of annealed aluminium is lower than 1350–H19, the rated strength of ACSS [12] is reduced by an amount dependent on the stranding compared to similar constructions of ACSR. In fact, a 45/7 ACSS conductor

(45 aluminium strands and 7 steel strands), with standard strength steel core wire has about the same rated breaking strength as conventional all aluminium conductors made with hard drawn aluminium wire. Similarly, given the low tension in the aluminium strands, ACSS does not creep under everyday tension loading. ACSS/TW constructions behave in the same manner as ACSS but have the added advantages [13] of reduced ice and wind loading and reduced wind drag per unit aluminium area.

9.11 References

1 Cigré SCB2 WG12: 'Conductors for uprating of overhead lines'. November, 2003
2 SAITO, T. *et al.*: 'Spiral-elliptic conductor with low drag coefficient'. IEEE Power Engineering Society Winter Meeting, Singapore, January 2000, **4**, pp. 2397–2402
3 GAUDRY, M., CHORE, F., HARDY, C., and GHANNOUM, E.: 'Increasing the ampacity of overhead lines using homogeneous compact conductors'. Paper 22-201, Cigré session, Paris, 1998
4 COUNESON, P. *et al.*: 'Improving the performance of existing high-voltage overhead lines by using compact phase and ground conductors'. Paper 22-209, Cigré session, Paris, 1998
5 LIVINGSTON, A. E.: 'Self-damping conductors for the control of aeolian vibration of transmission lines'. CEA paper 70-TR-225, presented October 1969, Calgary, Alberta, Canada
6 McCULLOCH, A. R. *et al.*: 'Ten years of progress with self-damping conductor', *IEEE Trans. Power Appar. Syst.*, May/June 1980, **PAS-99**(3), pp. 998–1011
7 ROCHE, J. B. and DOUGLASS, D. A.: 'T2 wind motion resistant conductor', IEEE Paper No. 84 WM 203-5, T-PAS, October 1985, **PAS-104**(10), pp. 2879–2887
8 KOTAKA, S. *et al.*: 'Applications of gap-type small-sag conductors for overhead transmission lines', *SEI Technical Review*, June 2000, (50)
9 SASAKI, S. *et al.*: 'ZTACIR-new extra-heat resistant galvanized invar-reinforced aluminium alloy conductor', *Sumitomo Electric Technical Review*, 1985, (24)
10 ASTM B856-95: 'Standard specification for concentric-lay-stranded aluminium conductor, coated steel supported (ACSS)'
11 ASTM B857-95: 'Standard specification for shaped wire compact concentric-lay-stranded aluminium conductors, coated steel supported (ACSS/TW)'
12 ADAMS, H. W.: 'Steel supported aluminum conductors (SSAC) for overhead transmission lines'. Presented at the IEEE PES Winter Power Meeting, 1974, Paper T74054-3
13 THRASH, F. R.: 'ACSS/TW – an improved conductor for upgrading existing lines or new construction'. IEEE T&D conference, New Orleans, LA, April 11–16, 1999

Chapter 10
Line construction

10.1 Introduction

Wood pole overhead line construction has changed significantly over the past few years. Most of the changes have been brought about by the need to mechanise a lot of the work previously achieved by manual techniques. Due to the downsizing of the workforce, previous methods are no longer practical. Line teams now need all the latest equipment in order to do the same amount of work previously undertaken by large teams of linesmen. The following chapter describes the construction of overhead lines in the order of work undertaken.

10.2 Site access

A great deal of damage can be caused by thoughtless access to private land. Compensation claims can sometimes amount to thousands of pounds, and goodwill for future access is completely lost.

A little planning prior to work commencing can alleviate any of the foreseeable problems likely to occur. Liaison with landowners is essential: they can give information about the particular access route that they would prefer to be used, they will also have first-hand knowledge of difficult terrain and areas where vehicles are likely to get bogged down. It is also valuable to discuss the type of vehicle access required. Overhead line construction can entail the use of various items of mechanical plant; winches, drum trailers, excavators and equipment trailers are commonly used for construction.

Specialised vehicles will help to solve difficult land access problems; low ground pressure vehicles are used to prevent vehicles becoming bogged down. However, in particularly bad areas, tracked vehicles, such as the Muskeg or Garron, are the best method of avoiding land damage.

All access routes need to be strictly adhered to and sometimes it is advisable to mark or fence the routes to be used, as the shortest route is not always the best.

Edges of fields adjacent to hedgerows tend to be drier and more able to sustain the weight of construction vehicles. Areas with a high concentration of peat soil are particularly difficult to access and require special consideration.

10.3 Excavations and foundations

In order for an overhead line to be successfully constructed, care must be taken to ensure that any excavation or foundation will be of sufficient strength and depth to support the structure and conductors. If the hole is too shallow the pole will not have sufficient stability; if it is too deep the clearance of conductors to ground level, roads, special crossings etc. will not be enough to comply with statutory regulations.

Poor conditions such as peat bogs and areas with a high water table need special foundations to be installed below ground level. These vary from a single wood baulk bolted directly to the pole side to a complicated arrangement of struts and timber baulks arranged to form a cantilever-type platform from which the pole weight is evenly distributed.

Excavations for overhead line wood poles were traditionally achieved by either hand digging or by mechanical excavator. The shape of the excavation was such that it allowed the pole to be reared or tilted into the hole. If manual digging techniques were used the hole would be stepped so as to minimise the amount of soil to be excavated. However, for all manual excavations some form of shuttering is required, as the minimum depth of excavation for the smallest wood pole is 1.5 m.

Both the above methods of excavation are time consuming and cause considerable land damage. However, recent moves within the industry have been to use power augers wherever possible to drill the pole hole. These large diameter augers are usually quicker than other methods and have the advantage of leaving the pole, once installed, surrounded by virgin soil.

Another problem frequently encountered, particularly in the Lakeland regions, is the need to excavate in rock. Explosives may be the only answer for really hard rock. Augers may be used for softer types.

Other types of excavation are needed for the installation of stay blocks and for erecting H poles. Stays require the excavation to be underpinned to allow the stay block to be positioned at 90° to stay the rod. They also need a slot to allow the stay rod to lie in line with the pole top. This means that the linesman needs to enter the excavation to underpin the bank. Shuttering of such an excavation is not easy, as the stay rod and stay block need to be installed together.

10.4 Pole dressing and erection

10.4.1 General

Dressing poles before erection is the easiest way to build an overhead line. Steelwork, stay tops, insulators and other fittings can be fitted easily while the pole is supported on a trestle at ground level. HV steelwork weighs up to 50 kg for a light-duty terminal cross-arm. The equivalent cross-arm for heavy-duty construction weighs up to 120 kg, so it is easy to imagine the difficulty of handling these weights at the pole top once it has been erected.

Poles are dressed in accordance with the specification requirement. M20 bolts are generally used for HV and M16 bolts for LV. Each bolt should have a washer behind the nut and, where possible, should be positioned so that the potential shear forces act on the shank of the bolt rather than the threaded part. Care is needed when fitting stay tops, as these should be positioned so that the weight is taken on the pole and not on the tie straps of the cross-arm. One strand is taken out of the pole top stay fitting to be bonded to the steelwork (only on HV). This is known as the king wire.

H poles require a large excavation to accommodate the underground bracing baulks and stability blocks. They can be up to 3 m long and usually 1.8 m deep. This also provides a challenge when shuttering as the linesman needs to access the bottom of the excavation to provide a firm and level base for the pole to rest on.

There are many ways of erecting poles and a lot depends on the site location and the access arrangements as to which method is used. The following is a list of methods used in United Utilities.

10.4.2 Pole pikes

Although they are seldom used these days, pole pikes may be required to manoeuvre poles into place when the position of the pole means that other methods are impractical. LV lines running at the rear of housing estates could well be good examples of the sort of area where this method could be used. It relies on good co-ordination of team members and care must be taken when moving pikes. The pole is simply pushed into the upright position by means of the pikes.

10.4.3 Spar holm derrick

Spar holm derricks are used to raise the pole partially to a position where it is possible to push or winch the pole to the vertical. Easily carried to site, they are useful where mechanical plant cannot gain access. A small hand winch on the side provides the lifting mechanism. Spar holm derricks can be used in tandem to lift large poles and side by side to lift H poles.

10.4.4 Falling derrick

The falling derrick is used in conjunction with a winch to provide the initial lift of the pole from the vertical. Once the pole has started to rise the derrick falls forwards and as soon as the fixing position of the winch rope rises above the top of the derrick it releases the derrick, which then has no further purpose.

10.4.5 JCB strimech

A turntable attachment to the front loader of a JCB digger, known as a JCB strimech, allows the operator to drive forwards while raising and tilting the pole into the prepared hole. Once in the right position it can be lowered into place.

10.4.6 Ford rotaclaw

The rotaclaw works on the backactor of the digger. This device grips the pole and a turntable allows the pole to be rotated through 90°. The pole is then swung into position and lowered into the hole.

10.4.7 Massey Fergusson jib

A simple device attached to the front forks of the digger, the Massey Fergusson jib allows the pole to be slung from the hook attachment with a sling. It is then lifted over the hole and lowered into place.

10.4.8 Marooka/lorry-mounted crane

A pole manipulator on the lorry-mounted crane is used to lift and position the pole.

10.4.9 Helicopter

A helicopter is an expensive way of erecting poles, but in some locations it can be the only practical way of doing the job within the time constraints. This method becomes more economical if used for erecting large numbers of poles. It has been used in this manner when the pole holes have been previously excavated. The poles are lifted with guy ropes attached, lowered into the holes and guyed off to ground anchors. Backfilling can then be accomplished when the helicopter has left the site.

A lot of distribution network operators (DNOs) are now looking at alternative ways of pole erection using live techniques. Mechanical plant used for these operations must comply with safety rules and only skilled workers are allowed to take part in these operations.

Backfilling of holes is an important phase of the work. All backfill material should be regularly consolidated (every 150 mm) and it is good practice to replace backfill materials in reverse order of excavation. Topsoil must be used to finish the job and this is usually left proud to allow for settlement.

Site tidiness is worth a mention at this point as this can cause trouble between landowners and DNOs. Any materials left on site will inevitably cause problems either to livestock, machinery or to members of the public.

10.5 Staywork

10.5.1 General

To provide stability to terminal and angle poles it is necessary to install stay wires from the pole top to ground anchors. Various configurations for stays have been used over the years and different sizes of stay wire were used until recently.

A typical stay would include the following components:

- stay wire – 7/4.00 mm strands
- helically preformed fittings
- stay insulator
- stay rods and blocks.

10.5.2 Stay wire

The purpose of stays is to provide stability to the line supports. Care needs to be taken in siting the stay in relation to the poles. A stay sited too close to the pole will produce what is known as a crippling effect on the pole. The smaller the stay top angle the larger will be the crippling force on the pole. Ideally, the stay top angle should be 45°.

At present, only one size of stay wire is commonly used for all construction types. This is 7/4.00 mm stranded with a minimum failing load of 101 kN. Because this is high tensile steel, it is not possible to splice the stay wire to the pole or the fittings, therefore helically preformed type fittings have to be used for any joints in the stay.

10.5.3 Stay insulators

Stay insulators are installed in all stays, regardless of whether the conductors are insulated or bare. Two types of insulator are in common use: type 1 insulators are used for LV, 11 and 33 kV earthed construction; type 2 insulators are used for 33 kV unearthed construction (all to ENATS 43-91).

Positioning of the stay insulator is also critical, as this needs to comply with statutory regulations as stated in the Electricity Supply Regulations 1988. It must be positioned no lower than 3 m in either a fixed or broken wire condition and no higher than would allow any jumper or dropper to touch below the insulator, should that jumper or dropper break loose at any point.

10.5.4 Stay anchors

Stay anchors are of various types, depending on the ground conditions and the mechanical load on the line. The preferred anchor is the augered type, which is screwed into the ground using the same driving unit as the wood pole auger. A big advantage of this type of anchor is that it is possible to proof load the anchors at the same time as they are installed. By having a gauge fitted to the hydraulics of the driving unit it is possible to convert hydraulic pressure into a figure for holding capacity.

For lines of light construction, load locked anchors are easy to install and have good holding capacity. These are driven into the ground by a hydraulic hammer and then pulled back to load lock the device by turning the anchor plate through 90°. Proof loading is carried out at the same time as the anchor is locked in position. A load of 35 kN is applied to achieve this.

In areas where neither of the above methods is acceptable, the answer would be to install a stay rod and block. This consists of a wood or concrete baulk through which a stay rod is fixed to terminate above ground level. Although this method has been used on most of the overhead lines in this country, it is now considered to be the least favourable option when installing stay anchors.

In some areas, particularly in the Lakes region, a good percentage of the land has only a thin layer of topsoil, under which lies rock. For stay anchors in this type of ground a special type of expanding rod is used. A suitable hole is drilled into the rock and the bolt screwed into this until it grips the sides of the hole.

10.5.5 Helically preformed fittings

The stay wire can be jointed to the pole top, insulators and stay anchors using helically preformed fittings. They are easy to use and can be adjusted by re-application if necessary (once only). They are simply wrapped around the stay wire and when tension is applied they tighten onto the wire to provide a firm hold.

10.6 Conduction, erection and tensioning

10.6.1 Main elements

There are four main elements for conductor erection. These are:

1 work planning
2 running out conductors
3 pulling up sagging and tensioning
4 making off and terminating.

Because of the different circumstances encountered on overhead line work, it is not possible to define a single method to cover all aspects. However, all jobs will fall under the general guidance notes associated with the above four elements.

10.6.2 Work planning

Prior to running out conductors, a survey of site conditions needs to be thoroughly undertaken. Hazards need to be identified and plans made to safeguard against any problem arising during stringing of the conductors. Typical hazards could include:

- rough terrain
- poor ground conditions
- trees
- roads
- railways
- canals, rivers and reservoirs
- buildings
- overhead lines
- school recreation areas.

This list is by no means complete, but it gives an indication of the types of problem that exist.

Wayleave constraints are another problem and these need careful attention. Agreements made with landowners need to be stressed to line teams in order to avoid misunderstanding.

Having assessed the physical obstacles that need action, the next stage would be to collect all the materials required. These should be assessed from the materials schedule for that line. Depending on the length of line and types of conductor, the manpower requirements can then be resourced.

Before conductor running begins, any work involving tree clearance or crossing protection (scaffold) should be undertaken. Farmers should be informed so they can remove livestock from the fields.

Final preparation would be to position temporary backstays at the ends of the section to be strung (not required on terminal poles).

10.6.3 Running out

In order to pull the conductors through the length of the line, special winches are used. For three-phase HV lines, a three-drum rope winch is usually used. The ropes are pulled through the section by hand and at the same time they are positioned in rollers at the top of the intermediate poles. It is possible to run through section poles by installing heavy blocks at the pole top and, with careful planning, long sections can be pulled out.

At the opposite end of the line to the winch, conductor drums are mounted in cradles with braking devices attached. These prevent overrun when pulling out and keep the conductors under light tension during stringing, which helps to keep them clear of obstacles. The winch ropes are attached to the conductors using conductor stockings and swivels. Tension is taken within the winch, the drum brakes are adjusted and then the conductor is pulled through to the winch end of the line. During the pulling in operation and subsequent sagging and making off it is essential that good communication links are set up and that all areas with public access are adequately protected.

10.6.4 Pulling up, sagging and tensioning

The conductors are then backhung, that is to say, they are terminated at one end of the section. Various methods of termination are used. These are as follows:

- helically preformed
- compression
- mechanical (bolted)
- wedge
- snail.

Sagging of the conductors is the next stage. Three things need to be known to work out the correct sag:

1 conductor size
2 span length
3 temperature.

The sagging chart for the relevant conductor/construction is used to determine the sag with reference to the span length and temperature. Getting the sag right is important, as any variation would mean either that the ground clearance would be infringed or the structures and conductor are overtensioned.

For new conductors an erection sag chart is used. This overtensions the conductors to allow for stretching. Some conductors are tensioned for one hour then made off.

A linesman can measure the sag of conductors in one of two ways: first, by use of sag boards nailed to the poles at a measured distance below the conductors, the linesman sights through the boards while the conductor is pulled up. When the dip in the conductor is level with the top of the boards the sag is correct. The second method is to use a dynamometer to measure the tension in the conductor. This is usually positioned on the middle conductor and the rest are pulled up to match and sighted in by a linesman standing to the side of the line.

10.6.5 Making off

Once the conductors are sagged they can be terminated by one of the two methods in section 10.6.4. This needs careful attention to accuracy, especially on single span sections, as any discrepancy at the termination will be multiplied tenfold in mid span. However, before finally terminating the conductors, the pole at either end of the section needs to be checked for plumb, and stays adjusted accordingly. Temporary backstays and conductors all need to be adjusted until everything is satisfactory. Most terminations are easily achieved at the pole top. However, fittings for ACSR conductors are more easily accomplished at ground level.

Once the conductors have been terminated, the intermediate positions can then be bound in or made with helical fittings. Again, checks must be made at the pole position to judge whether the pole is plumb both longitudinally and transversely. The cross-arms on HV lines also need to be checked for squareness with the line.

Binding in of the intermediate poles completes the conductor stringing. Section poles can now be jumpered through and all construction equipment removed. Re-instatement of pole positions and access routes can now take place, where no further work is to be required.

10.7 Plant installation

Various items of plant are installed on overhead line wood poles. On HV lines, the most common item is the pole-mounted transformer. These vary in size and weight from 15 kVA weighing only 200 kg to a 200 kVA weighing approximately 1.5 tonnes. Other items of plant could range from any of the following:

- auto-recloser
- auto-sectionaliser

- line switch
- drop-out expulsion fuses
- cable termination
- surge arresters
- fault indicators.

Recent developments in tele-control could mean that voltage transformers, solar panels and radio antennae may also be fixed to the pole.

10.8 Notices and ACDs

Protection against unauthorised climbing is a major concern when finishing off an overhead line. Few members of the public are aware of the dangers associated with wood pole overhead lines. Quite a few people refer to them as telegraph poles, believing that they have something to do with the telecommunication business rather than the distribution of high-voltage electricity.

To make people aware of the danger, notices are attached to poles to indicate that they are dangerous. All HV wood poles have got to have at least one safety sign ('danger of death') notice, generally to ENATS 43-90. In United Utilities it was decided to place two of these notices on every wood pole, regardless of whether it carried HV or LV conductors. Anybody now approaching the pole from any direction would be able to see the notices clearly.

All supports are numbered to aid identification in the event of a fault or some other form of work. This also helps when doing routine patrols. Other signs on the pole would usually be fitted to transformer and switch poles. These would be 'property' and 'nameplates'. These aid the identification of the system both to staff and to members of the public.

No other notices are allowed on the pole. However, in recreational areas crossed by overhead lines, it is good practice to position additional notices beneath the conductors to warn of their presence.

In addition to the warning signs, anticlimbing devices need to be fitted to poles and stays if there is any foreseeable risk of anybody being able to gain unauthorised access. Depending on the location, these ACDs can consist of a simple wrap of twelve turns of barbed wire or an outrigger bracket, either prewrapped with barbed wire or wrapped onsite.

In areas of high risk, enhanced ACDs are fitted. These consist of both of the above methods and constitute a considerable deterrent to any attempt at climbing the structure. All stays are wrapped with barbed wire to prevent someone trying to pull himself or herself up the stay wire and into danger.

Drilling of the pole to accept the plant needs care to ensure that the finished product is installed to be in line with the structure and that it sits level on the supporting steelwork. Wherever possible, the head of the bolt should be used to support the weight of the plant.

10.9 Connection to the line and to the earthing system

Connections to the line and to the earthing system are made when all fixings are secure. Line connections are sometimes made to bails using live line taps. Earthing connections are made to all HV steelwork and tanks of transformers etc.

Standard earthing arrangements for HV earth mats now consist of buried conductor of at least 70 mm^2 connected to three earth rods situated at the points of a 6 m triangle.

On live LV lines, voltage regulators and static balancers are quite common, although the installation of such items is becoming less frequent as the HV system is re-inforced.

The design of the poles must take account of the extra loading imposed by the fitting of plant, as some plant items might require an extra pole to support the additional weight. The general rule adhered to would be to install items up to 750 kg on a single pole and anything above to be installed on an H-pole structure.

10.10 Further reading

MORECOMBE, W.: 'Overhead power lines' (Chapman and Hall)

SMITH, S.: 'Study of overhead distribution lines and their design parameters' (Energy Networks Association)

ENATS 43-90: 'Anti climbing devices and safety signs for HV lines up to and including 400 kV'

ENATS 43-91: 'Stay strands and stay fittings for overhead lines'

Chapter 11
Inspection techniques

11.1 Introduction

The Electricity Supply Regulations 1988 (Part V, Paragraph 24) state: 'The supplier shall take all reasonably practicable steps to inspect his installations to ensure compliance with these Regulations'.

The electricity utilities (or DNOs as they are now known in the UK) therefore have a responsibility to inspect their networks and, in the context of this book in particular, their overhead lines. This is not only essential to maintain supplies in a costeffective manner and to be certain that regulatory clearances etc. are complied with, but also to provide a legal defence in any litigation where third parties may try to blame the DNO for incidents involving electricity supply lines.

Unfortunately, the regulations give no guidance on methods or frequency of inspections and so it is up to the DNO to decide upon a regime that will deliver the required network performance in terms of supply and safety.

The Electricity Supply Regulations 1988 (Part V, Paragraph 17) also state:

> All supplier's works shall be sufficient for the purposes and for the circumstances in which they are used & so constructed, installed, protected (both electrically & mechanically), used & maintained as to prevent danger or interruption of supply so far as is reasonably practicable.

So, not only does the DNO have to inspect their lines, but also maintain them in a manner to ensure a safe and secure supply. Again – no indications of how this is to be done are given.

Chapters 12, 13 and 14 will look in detail at inspection regimes, condition assessment and data collection and failure modes. This chapter will concentrate on the basic methods of inspection techniques.

11.2 Maintenance strategy

A maintenance strategy has to incorporate various areas:

- safety
- integrity
- technical procedures
- management systems.

The aim is to have a system that is sufficiently simple and straightforward so that it can be completed quickly and efficiently but that at the same time delivers all the condition and safety information that is required. It must also collect this in a manner that, these days, will interact easily with the DNOs data management system.

11.3 Inspection and maintenance routines of the past

Regimes used over the past century for OHL inspection and maintenance have been time based. The aim was to obtain a constant amount of information on the basis of which to take remedial action. Any defects found could be classified as urgent – requiring immediate attention – or not urgent – could wait until next planned outage for attention. This system therefore generated a steady pool of work and left the OHL network in a reasonably permanent state of good repair. It was, however, an expensive system that could in fact generate its own problems.

It can be assumed that each DNO has at some time implemented a policy that provides some form of routine inspection. However, the effectiveness of this policy in providing the necessary information to determine overhead line condition could be questionable. Data collected from overhead line inspections has generally been associated with the individual ability to recognise defects. The defects are generally categorised and prioritised as:

- urgent – statutory or safety issues
- planned maintenance – remedial works to be carried out in order to remove the defect at a later date
- refurbishment – defects identified that have little significance to the operation and safety of the system and can be removed during organised remedial works of the network.

The old adage – if it isn't broken, don't fix it – can apply to regular maintenance. Any system that is performing smoothly can often be upset by interference from the maintenance itself. The bath-tub curve (Figure 11.1) shows that even new equipment has a high failure rate until the gremlins are sorted out and everything settles down. Disturbance by a maintenance operation can re-start this curve with a high failure rate until the whole system settles down again.

This curve not only describes the whole system but also individual components. Each set of components can have a different lifespan – conductors say 40 years, nameplates and ACDs say 15 years, poles say 80 years, transformers say 60 years etc. So the network actually has several sub-sets of bath-tub curves.

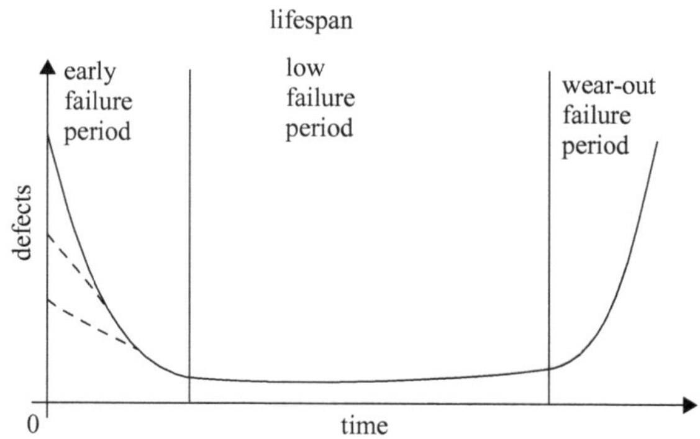

Figure 11.1 *The bath-tub curve*

11.4 Inspection and maintenance – current approach

11.4.1 General

The modern emphasis for electricity supply (laid out in the Electricity Supply Quality and Continuity Regulations (ESQCR), 2002) is to concentrate on security and safety. Any inspection and maintenance programme has thus to contain this concept. In line with the comments in the previous section, the current approach can be based on:

- safety and security
- component lifespans
- bath-tub curve.

The emphasis is now therefore on individual components. In this aspect a component may not be a physical entity but a design feature. Certain components may be perfectly in order but inefficient or ineffective due to modern development or improved line designs to new standards. Examples of this could be:

- the use of polymeric rather than porcelain components, e.g. in surge arresters
- the elimination of old-type stay insulators
- the removal of cross-arm mounted surge arresters at OHL/cable junctions in favour of the cruciform system
- the removal of old-type 'lossy' transformers.

11.4.2 Component groups

It has already been mentioned that bath-tub curves could encompass different lifespans for different components and typically:

- curve – 60 years
- curve – 30 years
- curve – 15 years.

One way to use a system based on different groups is to base a maintenance period on a common cycle such as 15 years. Such a system could be based on:

- precommissioning and data collection
- safety and security
- maintenance.

Regular maintenance is based on a 15-year cycle and defects found within this period are noted for attention at the next cycle end – unless the defect affects safety and security. Major works are therefore restricted to a specific time period for each network section.

11.4.3 Data collection

Overloading with data can be a major problem and in many cases it is essential to restrict the data collected from the regular inspection processes to useful condition information which can be fed into a condition-based maintenance programme. This is developed in more detail in chapter 13.

Specifically, data relating to wood pole condition must answer several questions:

- how many are there in each condition category?
- is sampling possible?
- is there a wood problem?
- is there a design problem?
- what inspection details are required?
- what frequency of inspection?

In specific terms, the real importance of wood pole support structures and inspection and maintenance regimes for this component is reflected in chapter 12, a chapter devoted purely to wood poles.

11.4.4 Foot patrols and aerial inspections

The frequency of routine patrols or ground inspections has generally been determined through consideration of the historical performance of plant and equipment, and environmental issues associated with the area in which the overhead lines operate. Safety issues, such as the location and the public's vulnerability to the dangers that may exist, would generally influence the requirement for line inspections to be carried out on a more frequent basis.

The inspection frequency of overhead lines has generally varied throughout the UK. In particular, some DNOs have used helicopter patrolling as an alternative to foot patrols and increased the helicopter frequency in preference to the foot patrol.

All companies register the information given in some form or other, it is, however, unfortunate that the information gathered is sometimes misleading or indeed inaccurate and is often then ignored. This can occur through lack of experience or knowledge on behalf of the inspector or indeed misinterpretation of results from onsite diagnostic tests. It is therefore essential that inspection and subsequent analysis of data from the field is carried out by trained and experienced individuals and that the data are interpreted accurately so that the most economic corrective measures can be taken to improve, where necessary, the performance of the overhead line network.

Unfortunately, this can be difficult to achieve in practice as, following data collection, the subsequent analysis and then implementation of the necessary remedial action may be flawed. Maintenance tends to be the first budget considered for reduction, meaning that resources are then stretched to carry out the inspections necessary to meet the policy requirements. Experienced staff are sometimes no longer available to make skilled judgements during the inspection process. The difficulty is in establishing what the policy for inspection and condition assessment of overhead lines should be to match current requirements and also be economically viable to be carried out effectively.

Without the key information relating to the overhead line's current condition, decisions about maintenance or future development cannot be optimised to maximise the use of the existing assets. The task of accurately establishing the condition of a modern overhead line network is very complex, but is necessary to predict long-term maintenance programmes. Typical networks comprise numerous lines of varying ages, structure type, configuration and conductors, most of which have been maintained to differing standards. The UK offers many different types of environment that can dramatically alter the rate of deterioration of individual components. The replacement over time of components within the overhead line therefore provides a complete mix of new and old components, often ageing at different rates.

A general inspection methodology is usually employed by the DNO to consider statutory or safety aspects associated with the overhead line. However, to actually assess the condition of the overhead line relative to its long-term performance a much more detailed data collection and monitoring exercise has to be employed. The more detailed assessments are generally carried out on those circuits identified to be poor performers.

Typical line inspections, requiring detailed condition assessment, are generally carried out by foot patrols, but increasingly, due to the advantages of aerial data collection and significant technological developments, helicopter use is becoming widespread.

As an alternative method to ground inspections, helicopters can provide a means of performing a faster and less expensive inspection while, at the same time, still providing a satisfactory level of detail. Consequently, the use of helicopters by electricity companies for routinely inspecting their overhead HV distribution lines has increased in recent years.

A typical combined foot patrol and helicopter inspection scenario might be as Figure 11.2.

11.4.5 Aerial data acquisition and processing

Technologically advanced systems provide the following benefits:

- efficiency
- accuracy

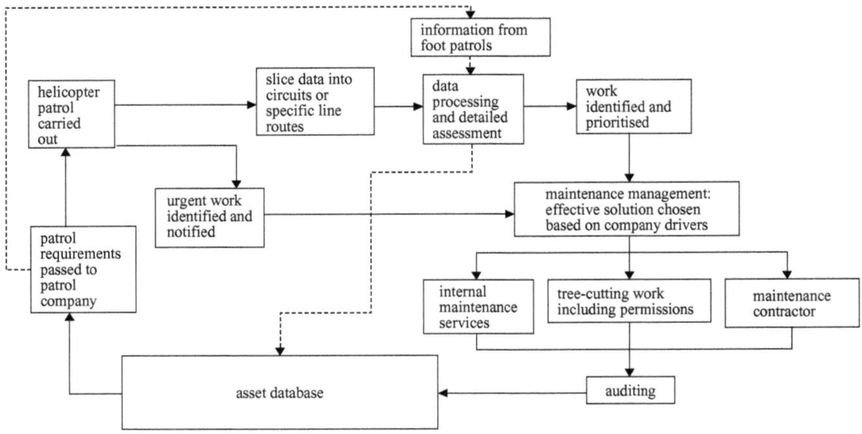

Figure 11.2 A typical schematic programme to combine foot and helicopter patrols and a network management programme

- auditable processes
- cost effectiveness.

Aerial data acquisition systems are available in many forms:

- observers
- aerial photography
- stabilised camera platforms
- Lidar
- strap down system (VidTracker).

Let us look at each of these in detail.

11.4.5.1 Observers

- generally experienced linesmen, engineers or operators
- without technical assistance the observer generally needs to be close to the line, often 20 m above ground level
- can be time consuming during flight to detail defects or non-standard components depending on the requirements of inspection
- limited methods available to spot GPS of assets
- low-level post processing to detail defects found during flights
- low to medium cost
- not auditable.

So there are problems with using an observer that include both personnel and cost scenarios.

11.4.5.2 Aerial photography

- high-altitude (300–6000 m) digital photography
- insufficient data to make condition-based decisions

- limited asset information
- limited methods available to spot GPS of assets
- post processing can be time consuming depending on the deliverables requested by the customer
- medium to high cost
- auditable process.

So, aerial photography has its advantages but the distances involved and the high cost render this system not always suitable for the distribution (as opposed to transmission) network.

11.4.5.3 Stabilised camera platform

- high-resolution digital imagery from 60 m+
- detailed information sufficient to make condition-based decisions
- asset information
- limited engineering data
- limited methods available to spot GPS of assets
- post processing is quicker than both aerial photography and Lidar processes as it is limited to purely interpretation of digital video
- medium to high cost
- auditable process.

Although more data can be obtained from higher resolution pictures of components, the cost factor again here weighs heavily. One good point is the ability to build up a library of photographs of equipment condition and environment that can aid in future maintenance and even litigation processes.

11.4.5.4 Lidar

- high-altitude (300 m) laser equipment recording highly detailed engineering information (survey data)
- digital video often operated during flight, however insufficient data to make condition-based decisions
- limited asset data
- limited methods available to spot GPS of assets
- post processing can be time consuming depending on the deliverables requested by the customer
- medium to high cost
- auditable process.

Lidar is based on a series of spot recordings from laser spot imaging. Data processing and analysis is time consuming and the whole process can be expensive.

11.4.5.5 Strap down

- cross-links high-resolution video and still images from 60 m
- detailed information sufficient to make condition-based decisions

- asset information
- limited engineering data
- post processing is quicker than both aerial photography and Lidar processes as it is limited to purely interpretation of digital video and stills
- GPS of assets and surrounding points of interest available with VidTracker data
- low to medium cost
- auditable process.

A strap down system combines still digital cameras linked to digital video systems that can be analysed in the DNO office. It can also link in with existing data management systems and all at a relatively low cost.

11.4.6 The key objectives

The key objectives for helicopter inspections are:

- GPS positioning on assets and/or tree locations
- vegetation survey and work programme
- ability to quickly prioritise tree-cutting requirements in relation to risk and deploy staff appropriately
- safety and security patrol
- auditable process from collection of data to implementation of work activities.

This allows the main purposes of inspection and maintenance regimes to be achieved:

- safety and security of line and components
- defects or faults identified
- natural ageing spotted, which introduces weaknesses that may lead to faults
- changes in environment that may lead to faults or danger to the public identified
- statutory requirements and/or policy implementation.

11.5 The choice to maintain or refurbish

11.5.1 Criteria for selecting circuits for refurbishment

Using statistical information relating to the various key parameters, a league table can be prepared to provide a basis for prioritising the selection of poor performing circuits. The four possible key areas considered could be:

- worst served customers
- interruptions/100 customers
- faults/100 km
- customer hours lost.

The combination of these four aspects of performance is then weighted with regard to importance and a league table of the worst circuits can be produced. Obviously, the key areas considered are based on fault data information received (via NAFIRS database in the UK) and the accuracy of the data will always only be as good as the reporting system adopted.

It is assumed that, statistically, the worst performing circuits can therefore be identified. However, proper analysis of the circuits identified at this stage should be carried out to ensure that the faults attributed to the circuit are those associated with the overhead line, and not, for example, a cable fault or termination failures of cable sections within the overhead line itself.

11.5.2 Technical assessment of the overhead line

Wood pole overhead lines are notoriously difficult to assess, especially old lines that have extremely varied condition states due to piecemeal maintenance and re-building work. Where detailed line condition can be provided it is, however, possible to extrapolate and analyse the data to provide a remaining engineering life, but this must reflect the design criteria on which both the condition and the design analysis is based.

In order to determine failure modes it is important to know the line, its components and how they perform. The technical assessment of a wood pole overhead line may, for example, consider the following factors:

1 The design of the overhead line
   - grade of poles
   - cross-arm width and type
   - fittings (insulators, pins, ties)
   - conductor
   - span length.
2 The topography of the overhead line route.
3 The meteorological information prior to and at the time of failure.
4 Historical data of the line performance.
5 Pole foundations.

The application of loads to the conductor is fundamental in understanding how overhead lines will perform in the field as it is the forces produced by these applied loads that will affect how the fittings' steelwork and structure perform.

11.5.3 Defect assessment

When categorising defects on overhead lines consideration should be given to the continued stability of the line and its safe operation. The inspection should consider each component inspected to be fit for purpose for a specified duration; should the inspection of the apparatus find that it falls below this requirement it should be noted as a defect.

The condition of an item of plant is its ability to perform its required function costeffectively, safely and to current specification. All lines should therefore be inspected to identify:

- weaknesses that may lead to faults
- changes in environment that may lead to faults or danger to the public
- deterioration of components
- failure to comply with statutory regulations or design standards.

11.6 Design data

The condition of the asset has always been the driving force behind the remedial actions that should take place in relation to overhead line refurbishment or re-build. However, what if the overhead line is in fact not fit for purpose in relation to its recognised geographical environment? This should also be considered as part of the condition assessment.

Experience has shown that where the design limits have been compromised the likelihood of failure during severe weather conditions will inevitably be increased. In order to assess accurately the existing overhead line's reliability level, data from line records and the overhead line survey inspections are analysed to determine whether they meet with minimum acceptable reliability levels.

The data that are required to calculate loads and clashing probability are:

- conductor type
- conductor phase spacing
- span length
- associated weather zone.

Required reliability levels are determined and then appropriate allowable span lengths can be calculated. This is the basis of the design health index described in chapter 13.

Using this methodology, the conductor clashing limits can be considered and steps taken to design out the potential conductor clashing with costeffective solutions that will improve network performance.

11.7 Data analysis

There are two main issues that have to be considered:

1. Natural asset life of overhead line components.
2. Likelihood to withstand extreme weather conditions (based on original design criteria).

A fit for purpose design must assume that the apparatus can provide an acceptable limit of both electrical and mechanical reliability, yet it is surprising how many overhead lines currently in operation do not meet with this requirement.

Using accurate, reliable data, however, can help evaluate all overhead lines and determine whether they do in fact meet with fit for purpose requirements. There are undoubtedly further issues that can be considered as part of the overhead line monitoring process, such as:

- statutory requirements (safety signs, clearance details, anticlimbing)
- equipment details (transformers, down leads, re-closers, fuses, lightning protection etc.)
- tree clearances.

A full condition monitoring specification and service will not only provide the user with important asset data on the condition of the overhead line network, it will also provide information on the strengths and weaknesses of the line in relation to extreme weather conditions specific to its design criteria. Once this information is available and a technical assessment of the information is carried out, costeffective recommendation can be provided, where necessary, to improve reliability and performance.

11.8 Business case

Any circuit considered for refurbishment must have a business case laid out. This will provide evidence detailing the existing overhead line reliability for its geographic environment together with the relevant information on the remnant life of the overhead line components.

The business case should detail the necessary recommendations to bring the overhead line to a fit for purpose criteria. This will include an investment appraisal of the overhead line options and reasons for proposing the recommendations submitted. It should also provide information relating to the expected new reliability levels, following implementation.

11.9 The modern approach

Condition assessment has led to a condition-based decision making (CBDM) process that has gained ground throughout the world. This has basically involved reducing data to a few specific condition categories, and, using fault data, prioritising maintenance and predicting component lifetimes. This process – now known as health indices – is explained in detail in chapter 13.

Chapter 12
Specific line inspection regimes

12.1 Introduction

Overhead line asset owners have a statutory obligation under the Electricity Supply Regulations to ensure that the network equipment is 'sufficient for the purposes for and the circumstances in which they are used' and 'so constructed, installed (both electrically and mechanically), used and maintained as to prevent danger and interruption of supply, so far as is reasonably practicable'.

In order to achieve this, inspection regimes have historically been introduced to consider the onsite condition of these assets and therefore determine what should be done to avoid both interruption of supply and danger to the public. Chapter 11 explained the techniques and philosophy of inspection; this chapter considers line component inspection in detail.

12.2 Field inspection of networks

12.2.1 Field inspections and data acquisition

This inspection is a detailed asset- and reliability-centred data acquisition programme that not only considers the existing and projected life of overhead line materials and apparatus, but also considers their design reliability to withstand severe weather conditions in their geographical locations.

This section is a brief look at the various aspects of line inspection and the necessity to collect relevant data, and it is intended to be used by inspectors for guidance only.

12.2.2 Guidance notes for distribution overhead line wood pole inspection

12.2.2.1 General

When inspecting overhead lines thought should be given to the continued stability of the line and its safe operation. The inspection should consider each component inspected to be fit for purpose for a duration not less than 15 years from the date of inspection.

12.2.2.2 Acceptable condition

Acceptable condition is the condition in which an item is able to perform its required function and/or meet the relevant specification. So all lines should be inspected with a view to identifying:

1 weaknesses that may lead to faults
2 changes in environment that may lead to faults or danger to the public
3 deterioration of components
4 failure to comply with statutory or design standards.

12.2.2.3 Safety and security inspection

The basic reason for inspection is to maintain safety and security so that the line can be safe and secure for many years ahead. Section 12.2.3 details the inspection requirements for the patrol. Supporting documentation (e.g. lists with inspection categories and pole numbers) is required to carry out these requirements efficiently and to enable the data to be recorded in a manner that allows an effective input to an asset management system.

The inspection should be carried out relative to the pole being inspected and the first span out from that pole to the next pole position, e.g. if the pole inspection sheet reads pole 1 this will also cover span 1 and relate to clearances and obstacles in proximity to span 1. If the sheet or data index reads pole 3 this will also relate to the third span out from pole 1.

12.2.3 Inspection requirements

12.2.3.1 Inspectors

The inspector must have appropriate experience and training to carry out the work requested, and the ability to refer to appropriate documentation to support their findings during the inspection. The inspector should also be identifiable for future reference in relation to any queries or clarification of results.

12.2.3.2 Date of inspection

The date at which the inspection was carried should be recorded relative to every span/pole inspected. It is essential that these are carried out as one item on the same date.

12.2.3.3 Line reference

The line reference can often be a mix of circuit reference relating to the feeder or source breaker and the transformer pole at the end of the line. Whatever format is used each pole must have its own unique reference, which will be made up of the pole number and circuit reference.

12.2.3.4 Voltage

There are a number of voltages that operate on the distribution network; this must be clarified prior to the patrol taking place as the apparatus on each of these overhead lines may vary specific to the voltage considered.

12.2.3.5 Conductor

Various basic aspects should be noted:

- span length
- type
- single phase
- three phase
- bare or covered.

In order to carry out a design analysis of the overhead line, each span should also be independently measured and recorded. Conductors should be inspected for signs of damaged stranding, evidence of clashing, birdcaging or evidence of fatigue during the inspection process.

12.2.3.6 Conductor type

Information relating to the conductor type is fundamental to the reliability performance analysis that is carried out following the data capture process. It is highly likely that this information can be determined by a foot patrol inspection. However, details of the conductor type and size should be found in existing line record files.

12.2.4 Structure details

12.2.4.1 Number ID

Some companies operate a unique number system for poles and have this numbering system applied to the pole itself. Some, on the other hand, will identify the pole merely by its pole position relative to the number on that spur or main line. Today almost every pole in the UK has a unique GPS (global positioning system) pole position that can be used as an identifier.

Whatever system is adopted, a number must be used to clearly identify the pole position. It may be that this ID will be a GPS for future reference together with the pole's current ID.

12.2.4.2 Pole grade

It is recognised that wood poles have historically been graded in the UK into three distinct categories: light, medium and stout. The poles are usually identified at the gauge mark 3 m from the butt of the pole, and this normally indicates the year of manufacture, pole length and manufacturer. Some wood poles may not have any identifying features, however, and the inspector should endeavour to assess the pole's grade relative to the circumference at the ground line:

1 light grade – <630 mm
2 medium grade – <785 mm
3 stout grade – >785 mm.

12.2.4.3 Structure type

Generally there are a number of variations in structure design/type relating to overhead line construction within the same line section. The inspector should indicate which structure type is therefore currently being inspected.

12.2.4.4 Structure usage

Each structure will be selected specifically in relation to its design requirements. There are generally five forms of structure design used in overhead line construction, and the appropriate structure should be noted. Where angled structures are identified the approximate angle of the structure should be noted. The angle of deviation is that relating to a continuation of one side of the overhead line and how much deviation from this imaginary line the overhead line takes (Figure 12.1).

12.2.4.5 Cross-arm spacing

If the design of the overhead line is known it is possible to gather the relevant information from appropriate specifications or associated drawings. If there is no information available relating to the phase-to-phase spacing relative to the cross-arm configuration then the inspector can approximate this by considering the distances from the centre of the pole to the cross-arm tie strap position and then from that point to the insulator position.

The centre of the pole to the cross-arm tie strap position on horizontal configurations has either been 380 mm for light construction (<35 mm^2 copper or equivalent) and 500 mm for heavy construction (>35 mm^2 copper or equivalent). The light duty tie strap at 380 mm spacing is generally a flat bar or flat bar twisted design. The heavy-duty tie strap at 500 mm spacing is normally fabricated from a steel angle section.

12.2.4.6 Cross-arm configuration

Cross-arm configurations have historically been of many different designs in the UK and abroad. Generally, the most common has been the horizontal configuration

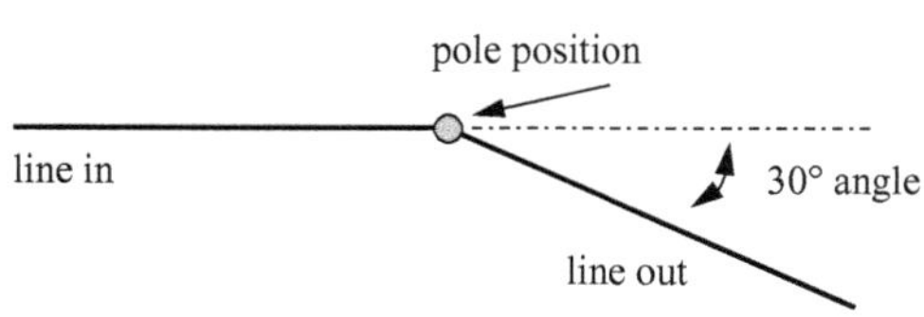

Figure 12.1 Angle of deviation

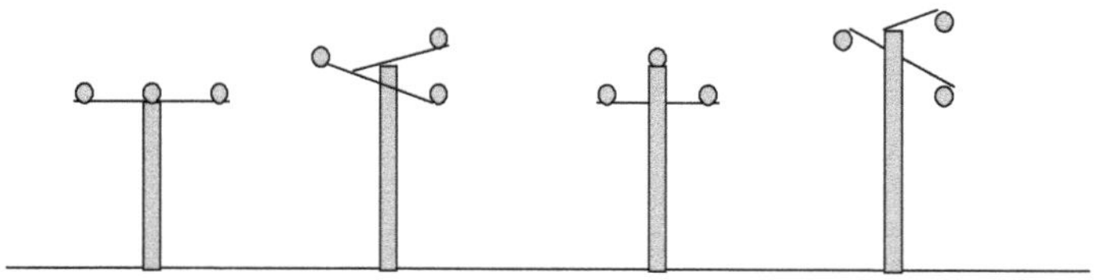

Figure 12.2 Typical cross-arm configurations

used in BS 1320, ENATS 43-10, ENATS 43-20 and more recently ENATS 43-40. There are, however, variations on this theme such as wishbone, delta and suspension (Figure 12.2). The inspector should identify which cross-arm configuration is present.

12.2.4.7 Number of stays

Note the number of stays identified at the structure installation. A stay is classified as slack if there is more than 2–3 in (50–75 mm) sideways movement in the wire. All stays should also have a porcelain or polymeric equivalent stay insulator or insulators inserted at a height of 3 m above ground (or a wall etc. from where a person may stand adjacent to the pole).

12.2.4.8 Stay spread

The distance from the pole at ground level to the point at which the stay enters the ground is the stay spread and should be noted so that the stay efficiency can be calculated.

12.2.4.9 Distance between stays

In order to provide suitable foundation capability stays should not be anchored closely together. Each stay requires approximately one metre of ground around it to pull against to hold its position in the ground. Otherwise, if two stays are trying to use the same ground (e.g. less than two metres apart where they enter the ground) the ground resistance will be weaker and the stay capability may be severely restricted. Data relating to the stay separation at ground level are therefore also required for stay efficiency calculations.

12.2.4.10 Leaning poles

If the structure is not plumb to the vertical axis then this can distort the crippling load on the pole as well as possibly lead to ground clearance problems. A leaning pole is therefore a defect and the amount of lean from the centre at the top of the structure to its true angle should be estimated.

12.2.4.11 Soundness

The capability of the structure to support its load and also possibly climbing linesmen can be critical. Test and inspections are therefore necessary to determine condition changes in structure strength and allow residual life to be predicted. Chapter 15 details wood pole decay mechanisms and remedial options.

12.2.4.12 Wood pole inspection requirements

It should never be assumed that a structure is safe to climb or can be used as a ladder support because of its good superficial appearance, as internal decay can be found anywhere on a pole. Any structure identified as being defective to the extent that it requires changing should therefore be labelled accordingly. One suggested procedure may be as follows:

1. All poles up to 30 years old to be visually inspected and hammer tested and all details recorded.
2. All poles over 30 years old to be subjected to a specific inspection process (ultrasound, Sibert drill or Mattson incremental borer etc.) and all details recorded.
3. Classification labels to be affixed, as appropriate, to poles >20 years old.
4. The labels may be 'A' (sound), 'S' (suspect) or 'D' (decayed) as required by the utility.

12.2.4.13 Visual inspection and hammer test

Check each pole for visible signs of decay (e.g. white fungus at ground level etc.) and hammer test between ground level and approximately 2 m up for audible signs of decay. It is usual to find decay starting on the side of the pole that faces the prevailing wind.

12.2.4.14 Hammer test

This consists of striking the pole a series of sharp but moderate blows with a wooden shafted hammer: 0.45–0.90 kg (1–2 lb) weight is recommended. Initially, the test should be applied at ground level and then around the pole until the height of normal reach is achieved. The hammer rebounding off the pole in a lively manner and producing a clear resonant sound gives an indication that the pole is sound. A defective pole is likely to give a dull rebound and a muffled sound or sometimes a hollow, drum-like note.

12.2.4.15 External decay

External decay is most likely to be found below ground level in older poles and in poles of low preservative retention levels. Extensive decay will result in the wood fibres becoming soft and spongy to the touch.

12.2.4.16 Prod test

In order to enable residual strength calculations, the depth of any surface rot is required. If spongy, fibrous material is identified on the pole surface then, using a suitable probe such as a thin bladed screwdriver, prod the pole. If penetration is detected to a depth of 50 mm at a particular test point or approximately 25 mm all around the pole circumference, the pole is unfit to climb and an appropriate inspection of the pole's strength is therefore required.

12.2.4.17 Residual strength value (RSV)

Residual strength calculation methods are detailed in chapter 15. The acceptable RSV will be decided by the utility but is typically 80 per cent of its original condition (the original figure is based on values from BS 1990 Part 1: 1984). Any poles falling below this residual strength figure may have to be condemned and therefore require replacement. In some cases a calculation of the actual pole load may be made and the pole left if it is still capable of carrying the load within agreed safety factors.

12.2.5 Insulators and connectors

Cracked or otherwise damaged insulators should be noted – and also insulators or connectors that are no longer recommended practice. These should be classed as defects.

12.2.6 Steelwork

The general condition of the overhead line steelwork should be inspected for reduction in steel content due to corrosion, visual distortion of the steelwork or evidence of fatigue. These may be due to industrial or coastal pollution, shock loading or general wear and tear, and should be assessed accordingly.

12.2.6.1 Auxiliary equipment

The presence of auxiliary equipment should be noted and its position relative to the stay insulator noted. Auxiliary equipment should be mounted such that the minimum height of exposed metal that is alive or may become alive at low or high voltage is 4.3 m. Power cables that are required to run between ground level and 3 m up the pole need a cable guard and its presence should be noted. Droppers may also require attention and their condition and type should be noted. Droppers should be installed moderately taut but should not exert any undue force on the main line conductors.

Transformers should have their visual condition noted – in particular for severe rusting, oil leaks, broken or cracked bushings, broken/damaged arc gaps etc. Particular regulations apply to LV connections, ACDs, AR, fusegear, ABS handles and inserts etc., and compliance or otherwise should be noted.

12.2.6.2 Surge arresters

The earth bonding of existing surge arresters should be noted and the presence of porcelain arresters in close proximity to the public should be noted as a defect.

12.2.6.3 Notices and nameplates

The presence and position of notices and nameplates should be noted as to whether they are in good condition and meet current statutory regulations.

12.2.6.4 Line clearance

Ground and other clearances should be checked for compliance with current statutory regulations and also local utility practice.

12.2.6.5 Change in land use (in proximity to overhead line)

Attention should be made to any new buildings or structures such as houses, garages, sheds, barns, aerials, that may have been erected which are not currently recorded on data sheets or supplied drawings or route plans.

12.3 Line maintenance strategies

The distribution overhead line wood pole population consists of poles which were installed under various pretreatment regimes and which will be approaching their end-of-life (EOL) situation at different ages. World-wide, utilities generally perceive that wood poles have an average life of around 40 years; this therefore indicates that a significant percentage of circuits continually require upgrading leading to a considerable investment in refurbishment programmes. Current policies on condition monitoring range from any pole showing rot to be removed, to poles with measured residual strength values or reduced factor of safety limit being selected for replacement. These policies have had a significant impact on the costs associated with refurbishment programmes as the interpretation of limited rot can range considerably depending on the inspector and devices used. A main characteristic of the decision making process is that the incipient decay will continue and reduce still further the pole's residual strength value.

The current costs associated with pole replacement are considerable. Depending on site conditions, pole usage, design requirements, labour and material the costs associated with a single pole replacement may vary from £450 to £1000 per installation (in the UK). Chapter 15 looks in detail at wood pole decay processes and remedial techniques that can help reduce pole replacement costs.

Chapter 13

Condition assessment and health indices

13.1 Introduction

Recent years have seen the derivation and population of health indices to rank assets on the basis of condition (in relation to proximity to end of life) as a means of assisting asset managers in formulating and justifying replacement or refurbishment plans. This chapter looks at the concept of health indices, which are based on utilising existing information, and summarises the experience gained over the past few years with developing and implementing health indices. This enables initial health indices to be derived and populated quickly, without the need for large-scale information collection. This gives a knowledge of present asset condition that can be immediately used to provide a very structured, well defined assessment of future condition assessment activities that can be used to focus and direct future asset management activities.

Overall, the continuing experience with health indices is very positive. Recent experience has demonstrated that they are of great value for distribution and transmission assets. As utilities attempt to develop a more integrated, risk-based approach to asset management there is a growing awareness of asset condition and how this relates to future performance. This will become an increasingly important component of replacement policies and plans in the future.

13.2 Health indices in context

Throughout the UK, and indeed many other parts of the western world, there was large-scale expansion in distribution networks and investment in electricity systems in the 1950s, 1960s and early 1970s. Since this time, the rate of asset replacement due to non-condition factors has been relatively slow. Consequently, the rate of replacement does not match the original rate of installation and the age profile continues to show an increase in average age for many asset types. As a result of this, many of the assets currently in use are approaching or have exceeded their original design life although they appear to be continuing to operate in a satisfactory, reliable and safe manner.

However, this raises a number of potentially significant issues, which include the risk of accelerating failure rate and the potential financial consequences of a significant increase in asset replacement spending.

The aim since privatisation, driven largely by regulatory pressure, has been to reduce short-term costs. This re-inforces concern about the long-term issues relating to end of life in the context of the ageing asset portfolios. Age in itself should not be a valid reason for replacing assets or justification for large increases in capital expenditure.

DNOs need to demonstrate that they can manage their ageing networks effectively and prevent serious engineering and financial consequences occurring as a result of increasing unreliability. There are two main risk areas:

1 the probability of failure
2 the consequences of failure.

A detailed understanding of condition provides a basis for a practical process to estimate future probabilities of failure.

13.3 Defining a health index

A health index (HI) is a means of combining varied and relatively complex condition information to give a result in the form of a single number. That number is intended to be a representation of the condition of the asset. It is important to emphasise that the objective of the health index is to define condition in terms of proximity to end of life or probability of failure. It is very important to understand the difference between condition that relates to end of life and condition that relates, for example, to maintenance. Traditionally, electricity companies have considered assets from the point of view of maintenance and in recent years much work has been undertaken to move towards condition or risk-based maintenance strategies. It is important to identify, and give appropriate weighting to, condition factors that are indicative of degradation that can ultimately lead to end of life, i.e. conditions that cannot be easily remedied by maintenance.

In order to end up with a single number it is necessary to first identify the significant condition criteria and then to define condition bands or condition ratings, which can be given a numerical code. Once information has been defined in this manner, weightings can be applied and a simple algorithm developed to generate a single number that applies to each asset and is a measure of its current condition.

A schematic representation of a health index is provided in Figure 13.1. In this case the health index is normalised on a scale of 0–10, the value of 10 indicating the worst possible condition. A score of 0 represents a condition in which there is no detectable deterioration. In reality, combining the scores for several condition criteria makes up individual health indices. In many cases incomplete information is available, such that scores can only be obtained for some of the condition criteria for individual assets. In this case the process of factorisation is applied. This means that

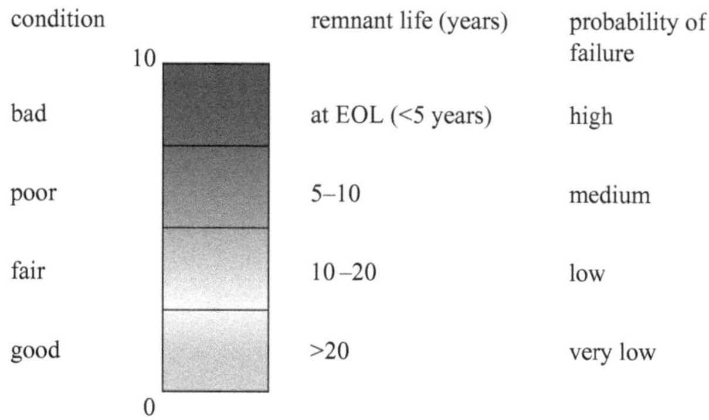

Figure 13.1 Schematic representation of a health index (courtesy EA Technology Ltd)

the health index is determined for the individual item by normalising the result for that asset on the basis of the information available.

An iterative process arrives at the final health index formulation. This involves an initial discussion to establish the significant condition criteria and definition of the condition ratings. The weightings, to reflect the relative importance of the different criteria, are then set and an algorithm is created to combine the information in a manner to give a representative indication of condition in relation to end of life. Once a reasonable volume of data has been processed, the resulting health index is compared back to the original condition information from which it was derived. This allows a calibration process to be undertaken to ensure that the relative values of health indices are representative of the conditions that have given rise to them.

In this manner the relative ranking of assets can be adjusted, as can the calibration of the health index, such that specific values can relate to defined conditions, i.e. end of life or to specific probability of failure. An index of 0–3 can be set to indicate very little significant degradation or increase in probability of failure. A health index of 3–5 can be indicative of significant deterioration with limited effect, at this stage, on the probability of failure. An index of 5 or above then indicates a significant increase in the probability of failure, which rapidly increases as the value moves from 5 to 10. This provides an opportunity to control the meaning of the health indices and link them firmly to the physical condition of the asset, based around an understanding of the degradation and failure processes. In this way the health index will have a real meaning in terms of condition and proximity to end of life.

13.4 The mechanics of the health index

To illustrate the formulation of a health index and the mechanics of using condition information to derive a single number, an example of a health index formulation for transformers is shown in Table 13.1. As this shows, eight condition criteria are used. These are criteria that can be obtained from specific oil tests, inspection results

Table 13.1 A transformer health index formulation

Condition criteria	Weighting	Condition ratings	Factors	Maximum score
DGA	4	1, 2, 3, 4 or 5	0, 1, 3, 7, 10	40
Standard oil tests	4	1, 2, 3 or 4	0, 3, 7, 10	40
Furfuraldehyde	5	1, 2, 3 or 4	0, 3, 7, 10	50
Doble	3	1 or 4	0, 3, 7, 10	30
Loading	3	1, 2, 3 or 4	0, 3, 7, 10	30
Tapchanger condition	4	1, 2, 3 or 4	0, 3, 7, 10	40
Bushing/ancillary condition	3	1, 2, 3 or 4	0, 3, 7, 10	30
Transformer general condition	4	1, 2, 3 or 4	0, 3, 7, 10	40

or background information. For each condition criteria up to four condition ratings (five for dissolved gas analysis (DGA)) are defined, ranging from condition rating 1 (CR1) which represents no deterioration to CR4 representing severe deterioration or high risk. Each rating is assigned a factor varying from 0 for CR1 to 10 for CR4. For each condition criteria a weighting of between 1 and 5 is assigned, based on the significance of the criteria to the end of life of the transformer. As can be seen, the condition of the paper insulation, as measured by the furfuraldehyde level in the oil, is considered to be the most significant individual criteria and hence attracts a rating of 5. Multiplying the condition rating factor by the weighting gives a score for each individual transformer for each condition criteria. The maximum score corresponding to a CR4 condition for each of the criteria is 300. This is the value that is used as the basis for normalisation when there is information for all nine condition criteria.

In the simplest form of a health index the individual scores are calculated for each criteria, added together and normalised against the maximum score onto a scale of 0–10 to give a health index. However, deriving a health index in a purely cumulative fashion may not adequately represent condition when the initial assessment is applied. Therefore, multiplication factors can be built into the algorithm which enables higher significance to be given to particular conditions and the overall combination to be calibrated to give the desired output.

13.5 General approach

When defining a condition rating for each condition criteria it is helpful to restrict the number of ratings to a relatively low number, e.g. 4. In this system four condition ratings are defined in general terms of what each condition means:

1 CR1 is a condition in which there is no detectable or measurable deterioration and no increased probability of failure.
2 CR2 is where there is evidence of deterioration that is considered to be normal ageing and has no significant effect on the probability of failure.

3 CR3 is a condition where there is significant deterioration that increases the probability of failure in the short to medium term.
4 CR4 represents severe degradation and indicates an immediate, significant increase in the probability of failure.

Even in a four-condition rating system, there does not have to be condition ratings for each condition criteria. There may be a pass/fail test, which would be set to CR1 or CR4. In other cases the information available may more easily be split into three categories that may be CR1, CR3 and CR4. Similarly, a three-case system could use CR1, CR3 and CR4 etc. This is a key element of developing and applying health indices – to take condition information and convert it into simple condition rating bands. This often needs very significant simplification of relatively complex data.

It is important to recognise that in many cases the information will not have been collected with the purpose of being used in this manner and therefore some flexibility and careful information management and manipulation will be necessary.

13.6 A practical approach to health indices

13.6.1 Sources of information

Health indices have to be based on utilising, whenever possible, existing information. Therefore, rather than attempting to define the perfect health index formulation for a given asset, it is necessary to consider what information is available and decide whether or not it is possible to derive a viable health index that will give a meaningful assessment of condition or ranking of assets on that basis. There is often sufficient information available to derive and populate a credible health index that will enable a useful definition and significant differentiation of condition. The formulation and population with existing information will then come up with an initial health index. The level, nature and quality of information available will determine how well differentiated the definition of condition is. For assets with significant quantities of information that relates directly to the degradation and failure processes that are most significant a high level of differentiation can be achieved. For asset groups with less detailed or less specific information a much less well differentiated definition of condition can be achieved.

Although the results can be used to assist in asset management decisions, the process will also help to clarify how further condition information can be most beneficially obtained. Plans can then be put in place to progressively improve the quality and differentiation of the health index by ongoing information gathering. The practical approach to developing health indices is described by the flow chart in Figure 13.2. As this shows, if when assessing the initial viability of the health index it is decided that there is not sufficient information available to formulate and populate even a basic health index, then decisions can be made how best and most economically to collect the necessary information. In some cases this may involve some targeted sampling and gathering of quite specific and detailed information for a relatively small number of the assets in the population. Alternatively, for more numerous, less significant

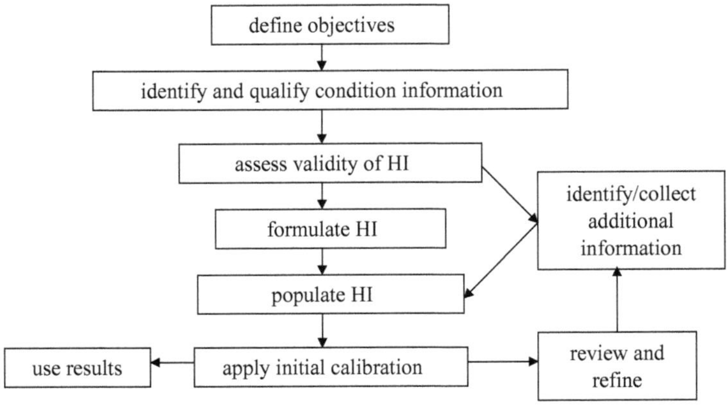

Figure 13.2 Representation of the pragmatic development of a health index (courtesy EA Technology Ltd)

assets where condition assessment is traditionally not considered viable, for example pole-mounted transformers, a programme of simple condition assessment can be constructed and implemented within normal inspection and maintenance activities to gradually build up the knowledge of the asset condition.

The great advantage of this approach is that it enables an appreciation and definition of condition to be achieved relatively quickly for many asset groups. It also defines the optimum ongoing information gathering requirements to allow effective and continuous improvement of asset condition definition.

13.6.2 Purpose

The purpose of the health indices discussed in this paper is to provide a consistent overall assessment of condition. It is recognised that there are other factors that will affect the ultimate decision for individual assets such as:

- criticality
- obsolescence
- safety issues
- environmental concerns
- etc.

However, in order to apply these appropriately a consistent appreciation of condition and an understanding of what that means in terms of future performance is essential.

13.6.3 Formulation

When formulating a health index there are two types of information that can be utilised. First, there is specific condition information that relates to individual assets, e.g. individual inspection reports for wood poles. However, for many distribution assets, particularly at the lower voltages, there will be relatively little specific condition information. This is sometimes thought to mean that it is impossible to achieve a practical and economic assessment of condition for such assets. Although there may not be much specific condition information for some assets there are almost always 'risk factors' which relate to condition, i.e. factors such as:

- reliability
- failure rates
- specific problems
- generic problems
- design issues
- environmental factors etc.

All of these can be used to prioritise or rank assets by risk of being in poor condition.

In many cases, for electricity distribution assets there is a wealth of such information available both within the parent company and within the wider technical community. There is also often a good understanding of the degradation and failure processes that enable condition and risk factors not only to rank assets but also to provide a basis for estimating remnant life. The more specific condition information that is available the more definitive the health index can be. If there is very little specific condition information and only some risk factors available the health index may still provide a useful means of prioritising assets. More detailed condition assessment activities can then be targeted in the most costeffective manner.

13.7 Risk factors

The use of risk factors is limited to those that relate to condition; non-condition-related risk factors (such as obsolescence or lack of functionality or criticality on the system) that relate to the consequences of the failure rather than the risk of failure are not used. It is very important to understand that the health index is an attempt to define the condition of the asset. It is recognised that the overall risk to the owner is a combination of the probability of failure and the consequences. Many companies are now developing risk models where the consequences of failure are assessed. The intention of the health index is to provide a means of understanding the probability of failure as an input to such models.

As most inspection processes to date have been designed to support and direct maintenance programmes, much of the available information is in the form of defect reports. Although this is not ideal for use in health indices, it can still be used. However, once utilities embark on developing health indices the benefits of adopting a fundamentally different information gathering process become clear. A positive reporting system using limited well-defined condition ratings enables information to be collected in a much more consistent and useful way. It also provides the basis for more efficient information storage, processing and retrieval systems. In most cases the value obtained from inspection information can be greatly increased, with the long-term costs of inspection and information gathering remaining broadly constant or in some cases being reduced.

13.8 Examples of health indices

13.8.1 General

To further illustrate many of the points made above and to consider the important criteria for different sub-station assets some examples of health indices that have been derived and populated with individual electricity companies are now presented and discussed. It must again be emphasised that this is not an attempt to define a perfect health index or indeed a health index that would be specifically recommended. These are all health indices that have been derived based on the pragmatic principles previously discussed, utilising information that was available within the company concerned. Experience to date has been that in every case the final details of the health index formulation have been different for similar assets in each company.

13.8.2 Distribution switchgear

It is clear that there are currently more active end of life definitions for distribution switchgear than any other asset group. These include obsolescence, generic problems, operational restrictions, poor reliability etc. Condition by itself is not widely used except as a qualifying factor within the other criteria in order to prioritise replacement.

When it comes to defining a health index for distribution switchgear it is likely that there will be limited specific information available initially. However, it is quite possible to define a very reasonable health index based on information that can be collected during routine sub-station inspections and maintenance. Consideration of the criteria used in this health index indicates that some of the information relating to the internal condition can only be obtainable at maintenance and therefore it will take many years to build up a complete condition profile for the population. However, if the information is collected systematically at maintenance it will not be long before there are results for a significant sample, which can be used to estimate the overall condition of the population. The same formulation can also be used as a consistent means of determining capability for any individual unit that is being considered for replacement, or to compare the condition of different sub-populations (e.g. different types).

When all of the condition information has been collected and a series of condition codes assigned to each piece of switchgear, there will be an overall condition score. This is then factorised to give a condition code on a scale of 0 to 10.

This formulation includes specific information, oil tests and internal condition that require invasive maintenance to obtain. However, as a short cut, the knowledge obtained from recent programmes such as the statistical sampling could be used to populate a risk factor to take account of internal condition. Indeed, there is sufficient knowledge of the degradation process, the susceptibility of different types and the effects of environment to derive a useful health index based mainly on risk factors.

13.9 Prioritisation

Having formulated and applied the health index to the available condition information and produced a ranking or prioritisation of the assets on the basis of proximity to end of life, some initial calibration can then be applied. As previously discussed, the form and detail of this calibration will depend on the nature of the assets, the level of condition information available to derive the health index and the requirements of the host electricity company. In most cases an attempt is made to define the range of health index values that constitute assets presently at end of life. Beyond this some attempt may be made to assign remnant life to the remaining assets. Alternatively, some indication of the requirements for ongoing management of the assets based on their overall condition, i.e.

- normal maintenance
- increased maintenance
- major overhaul
- replace

could be assigned. Alternatively, the health index could be used to define a risk of failure within a particular time frame.

In terms of remnant life and time frames used in conjunction with health indices, these are often in broad bands, e.g. five-year periods, particularly if the information is being used in the context of a regulatory submission.

13.10 Use of generic information

It is essential to appreciate that the health index is not designed to identify equipment that should be replaced for a non-condition reason. For example, a type of equipment deemed to be unfit for purpose due to a design or specification issue or obsolescence may be in very good condition and does not necessarily have an increased probability of failure. In such cases the generic issues should have a limited (possibly zero) influence on the health index. This can lead to asset managers being unhappy with the health index results of an asset group they have previously identified for replacement not indicating high risk.

In such cases it is important to recognise that the health index reflects risk-related asset condition and probability of failure and there are other potential risks that could lead to a decision to replace equipment. Assets to be replaced for non-condition reasons can be identified separately. Indeed, a simple non-condition EOL override factor can be defined. This is based on specifically identifying the non-condition criteria that could led to an EOL decision, and then scoring and combining to provide clear definition and a means of ranking types or groups potentially at EOL on grounds not directly related to condition.

The issues around the use of generic information for distribution equipment are relatively complex and have led to vigorous discussions with each company that has derived health indices. This has resulted in three distinctly different uses/approaches:

1 The generic information is built into the health index formulation, as described above.
2 The basic health index is derived purely from other (condition) information, and then a second value (an experience factor) is derived from generic experience; this is used to adjust the basic value. This leads to three values being displayed for each switch: a basic condition index, the experience factor and the combined health index. This approach is only useful where there is significant specific condition information available so that the condition of the switchgear can be reasonably assessed.
3 The switchgear is ranked or banded purely on generic considerations, the simple generic index is then modified by any available condition information. A company that had very little specific condition information available has used this. The starting point, the initial generic ranking, is essentially the traditional EOL process used for switchgear; as more condition information is collected this can be progressively modified leading to better discrimination related to condition and probability of failure.

13.11 Output from health indices

Some typical outputs from recently derived health indices are shown in Figures 13.3 and 13.4 (both courtesy of EA Technology Ltd) to illustrate the nature and potential value of the information that can be obtained. These include assets such as distribution

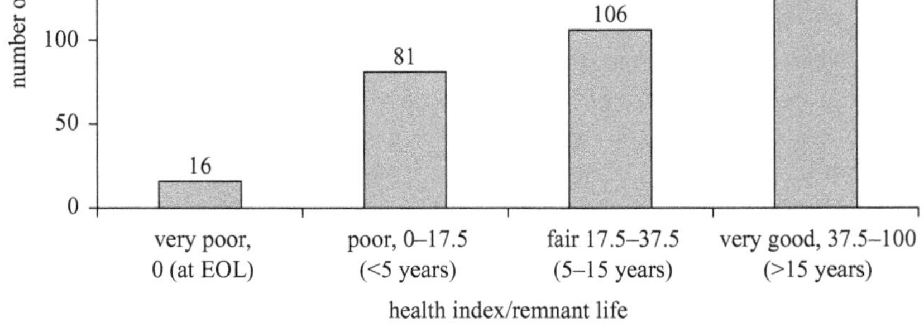

Figure 13.3 Phase conductors (courtesy EA Technology Ltd)

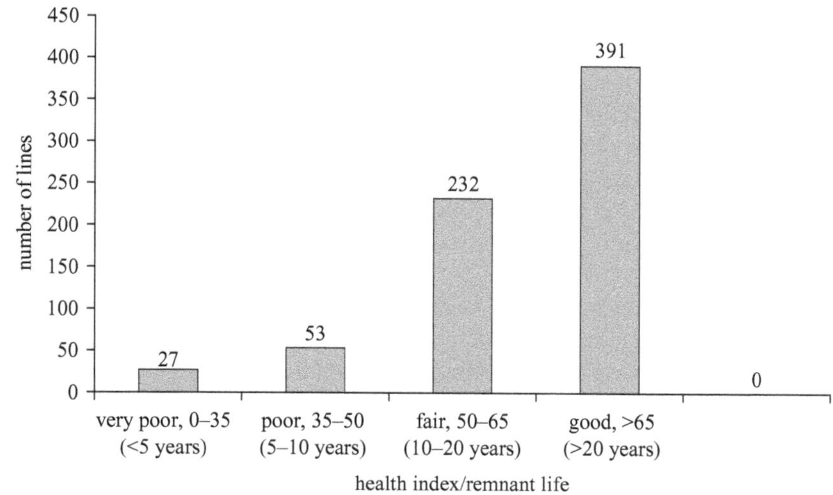

Figure 13.4 Distribution OHLs HIs and the need for refurbishment (courtesy EA Technology Ltd)

OHLs where the health index is a combination of specific information and condition-related risk factors. This information was provided in a recent CIRED paper [1].

13.12 Benefits of developing and implementing HIs

Developing and implementing health indices for the whole range of distribution assets enables a consistent use of engineering knowledge and experience to underpin ongoing asset management programmes. They are particularly valuable for mature and stable asset bases such as the distribution networks of many electricity companies.

Health indices provide:

1. A consistent means of utilising available condition information (and risk factors) to define proximity to end of life.
2. A structured process to define future condition assessment needs.
3. A means of utilising existing engineering knowledge and experience to predict future performance and failure rates.
4. A consistent, defined reference point to aid decision making.
5. A basis for planning and justifying future replacement/refurbishment plans.

13.13 Beyond a health index

The health index value for an individual asset, by definition, is related to its probability of failure. The greater the health index value, the higher the probability of failure. Recent work has resulted in the development of protocols to enable the health index to be translated into absolute values of probability of failure and to estimate the change in the health index values (and therefore probability of failure) into the future. Once a company has a consistent means of defining current condition and future performance of assets, the effects and consequences of any intervention strategies, replacement, refurbishment, repair etc. can be quantified.

These processes further emphasise the value and potential of the health index approach. The current and future probability of failure values can be used as an input to the risk models increasingly being developed to define and justify future investment plans.

The health index approach then becomes a vital component in an overall risk-based asset management process. The process of formulating and populating health indices also has another potentially very significant benefit. It forces companies to critically consider the information they collect and leads to a very clear appreciation of the value of changing the nature of information collected during routine activities. Traditionally, companies collect defect information, often by exception, to direct maintenance activities. By changing to relatively simple condition qualified assessment of equipment much greater value can be obtained with little or no increase in cost of collection. In companies that have adopted health indices the potential value of changing the nature of information collected, and maximising the opportunities to collect information during routine activities, has become obvious. This provides the justification and momentum to introduce radically different information collection regimes with enormous, long-term significance.

The change in HI with time can be estimated by using a rate of degradation for a process (e.g. wood pole decay, oil degradation etc.) and applying this to the HI as:

$$H_{t1} = H_{t0} \exp^{(B \cdot t1 - t0)}$$

where H_{t1} is the new (future) HI based on the HI value now at $t = 0\,(H_{t0})$. The factor, B, is obtained by estimating, for example, the time to degrade HI_5 to HI_{10} for the asset and checked by calculating lifetime range for the asset group. This knowledge of future asset condition can be used to plan maintenance/replacement regimes. For example, an estimate of future HI profiles and failure rates can be made for different maintenance/replacement scenarios. Note that for wood pole OHLs many weather-related faults are design related not condition related, so it can be useful to consider a separate design index.

13.14 The design index

To construct a design index some standard is needed to relate it to. In the case of wood pole lines, the process could be:

1. Compare actual line design with ENATS 43-40 design (or other current standard).
2. Use basic design information (span length, conductor size, weather zone etc.).
3. Normalise design index (as the HI) and match profile against design- (weather-) related faults.

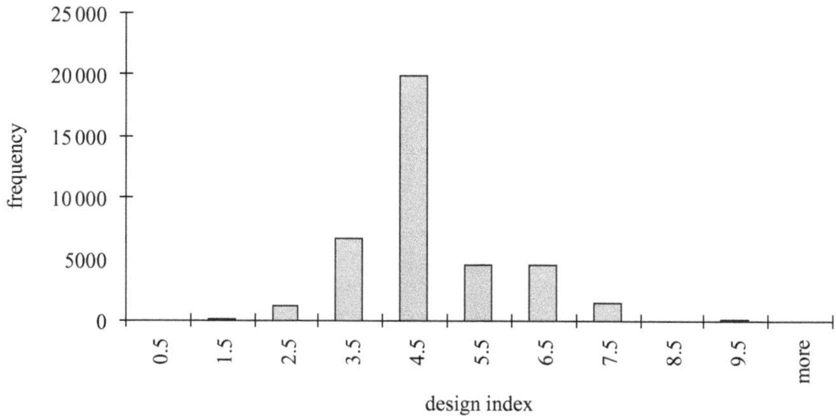

Figure 13.5 *Design index for 11 kV wood pole lines (courtesy EA Technology Ltd)*

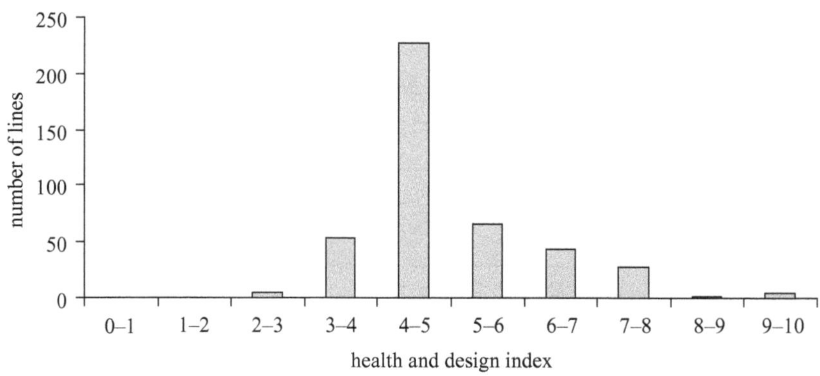

Figure 13.6 *Combined HI/design index for wood pole lines (courtesy EA Technology Ltd)*

Typically, a design index based on these principles for 11 kV wood pole lines in a utility may look like Figure 13.5.

This can be combined with an HI for the same lines, to produce a combined HI and design index (HI/DI), which gives a better overall picture of the asset condition. A typical HI/DI index for 11/6.6 kV lines is shown in Figure 13.6.

So the introduction of HIs into the design shows that some lines are in urgent need of replacement and in general the asset is in slightly worse condition than indicated by the design alone.

13.15 Future fault rate reduction

It is also possible now to introduce future condition analysis as discussed in section 13.13. Allowing degradation factors etc. according to known decay rates,

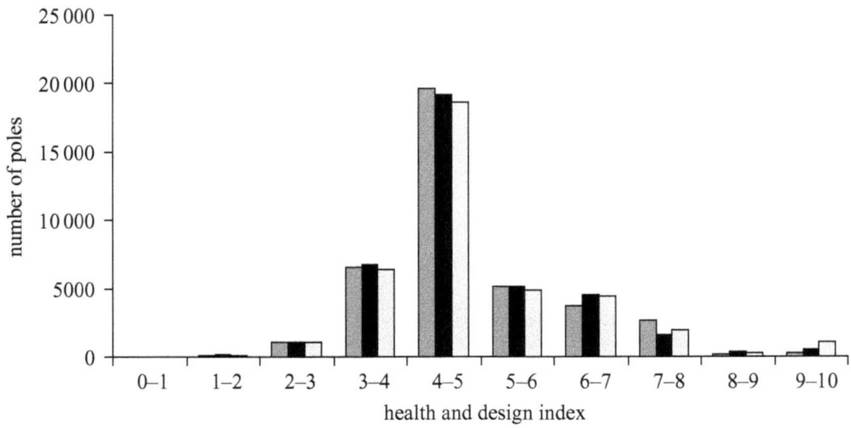

Figure 13.7 *Combined HI/design index at 0, 5 and 10 years (Courtesy EA Technology Ltd)*

Table 13.2 *HI/DI and future fault rate scenario (courtesy EA Technology Ltd)*

Year	HI/DI calibrated fault rate	Non-related HI/DI fault rate	Overall fault rate
0	0.0069	0.0028	0.0097
5	0.0075	0.0028	0.0103
10	0.0090	0.0028	0.0118

it is possible to re-draw Figure 13.6 for five and ten years in the future. The combined present and future index is shown in Figure 13.7.

To turn this to one practical purpose, the fault rate of a section of network can be estimated on the basis of the HI/DI. The fault rate needs to be correlated with Nafirs data initially and then split into what is affected by HI/DI (e.g. pole decay) and what is not (e.g. third-part damage). A table of future fault rates can then be drawn up and may typically look like Table 13.2.

Another way to look at this is to see the effect of various scenarios e.g. percentage re-build of the network section and see how the fault rate is affected. This can then be compared with other scenarios (e.g. re-conductoring, refurbishment, design change etc.). A typical comparison of re-build scenarios is shown in Table 13.3, using the data from Table 13.2.

Again, this data can be costed and the scenarios compared with alternative scenarios of fault reduction (e.g. improved response times).

Table 13.3 *Fault rate (year 10) with various re-build scenarios (courtesy EA Technology Ltd)*

% of lines re-built	HI/DI fault rate	Non-HI/DI fault rate	Overall fault rate
0	0.0090	0.0028	0.0118
4	0.0075	0.0028	0.0103
6.5	0.0069	0.0028	0.0097
15	0.0042	0.0028	0.0070

13.16 Conclusions

This chapter has attempted to illustrate the potential value to electricity companies of a structured use of condition information and in particular the derivation of health indices for major asset groups. Condition information can be used in a very constructive and positive manner to enable genuine condition-based management programmes. In the context of a heavily regulated industry as in the UK, with large numbers of relatively old assets, this type of approach has offered major opportunities to implement effective ongoing asset management. The combination of design indices as well can provide a realistic view of the asset and, when combined with fault records, the overall scenarios can be determined for various times in the future. This can allow specific policies to be evaluated on a cost/benefit basis.

13.17 Reference

1 HUGHES, D. T.: 'The use of health indices to determine end of life and estimate remnant life for distribution assets'. 17th CIRED, Barcelona, 12–15 May, 2003

Chapter 14
Failure modes in overhead lines

14.1 Introduction

14.1.1 The bath-tub curve

The bath-tub curve (see chapter 11) represents the simple lifecycle of almost any mechanical device. After early faults are corrected and the failures weeded out, there normally exists a period of stable performance.

Hence the curve – which represents the average performance of any overhead line component – starts off with a high fault level before dropping down to a steady operational low fault-level performance. This, simplistically, should carry on till the elements of old age creep in and things start to go wrong, fault levels increase and a time comes when a decision has to be made to put an end to its life before it becomes a total economic liability.

This is a simplistic view as the units may have maintenance – either regularly or when inspection demands. Each sub-maintenance period will have its own sub-bath-tub curve. At each inspection the decision has to be made as to the probable economic performance or lifetime expectancy of the unit to determine whether maintenance is necessary. Maintenance can extend life but it is not always economic. At some point a decision has to be made to run the unit to its natural end-of-lifecycle and then replace rather than maintain.

14.1.2 Environment

The particular lifecycle of any type of unit is dependent on the environment in which it operates and the workload it has to undergo. It is necessary, in order to obtain optimum economic life, to use units that are fit for purpose. The overhead line (OHL) must be built so that it can meet all the expected wind, snow/ice etc. effects throughout its life, i.e. the 50-year return period weather.

The protection part of the OHL system must also be designed to withstand the expected lightning intensity. If protection fails then the bath-tub curve will meet a premature end due to the interaction of the OHL components (e.g. surge arrester failure leading to transformer damage or poor design leading to conductor failure by clashing).

Apart from the individual incidents of storms, there is an insidious pattern of corrosion in the UK, especially on exposed steel or aluminium components, that requires protection or variable design considerations in polluted areas (e.g. within 10 km of the coast in the direction of the prevailing wind).

14.1.3 Workload

The OHL must withstand its expected workload and also the anticipated number of damage and non-damage faults that lead to switching overvoltages, lightning surges etc. On top of this, the line normally has a lifetime of 40 or 50 years, during which load demands will change due to urbanisation, development of green field sites and industrial decay. In order to match load and the current network facility, OHL components are often pushed to their perceived limits and sometimes temporarily beyond them. This will distort the lifecycle and lead to, for example, transformer overheating, regulatory line problems or switchgear faults.

14.2 Line components – failure modes

14.2.1 Asset performance

The extensive network construction in the 1950s and early 1960s means that now many overhead lines are reaching the end of their life. If a line needs refurbishment, is it best to replace only those parts that need replacing? Is there a need to replace those liable to be replaced in the next ten years? 20 years? Economic choices have to be made. In the case of underground cables the end cycle is often rapid and invisible, whereas the OHL conductor is visible and easily inspected.

In a similar vein, how much does future asset performance depend on the quality of installation work now? This is a relevant question these days with the increased use of contractors and turnkey projects due to the loss of specialised experienced personnel from the utilities.

An element not anticipated when many lines were built is the effect of regulatory changes. These can change in the short term, leading to asset management problems for equipment that was expected to last for decades but is suddenly no longer suitable. Typical examples are:

1 Cross-arm mounted surge arresters which have long earth connections and as such are very inefficient.
2 Old-fashioned stay insulators which need replacing with modern versions.
3 Design changes forced by new European standards.

Finally, the cost of holding multiple spares can be excessive and is another factor to be considered when the asset life of particular components is assessed.

14.2.2 OHL conductor

Conductor failure is not common. There may be, on average, 2000 conductor failures in the UK each year through storms or corrosion, out of a network of over 600 000 conductor kilometres. The main causes of HD copper conductor failure is normally overloading by snow/ice, tree problems and clashing (leading to arcing) in high winds. Copper is corrosion resistant and easily repaired, so long lifetimes can be expected. It also has a high vibration tension limit. The major problem with many small copper lines (using 16 mm^2 HD copper conductor) is lack of capacity leading to necessary re-conductoring.

Aluminium conductors (AAC, AAAC and ACSR) are subject to the same failure mechanisms as copper with the additional problems of corrosion near coasts and lower vibration tension limits, leading to fatigue failure. In many DNOs, however, the main cause of conductor failure is weakening due to crevice or galvanic corrosion.

Covered conductors can avoid the problem of clashing and reduce tree problems but have an increased susceptibility to vibration fatigue and lightning. At this stage it appears that corrosion can be avoided and suitable operational and design considerations can lead to reduced vibration and lightning problems.

14.2.3 Jumpers and stays

Too small a conductor size, vibration and poor compression joints can all cause jumper failures. Poor installation and the use of inappropriate connections can also lead to fault conditions. Stays are frequently broken at the pole connection, especially if eyebolts are used. Stays rarely break due to overloading, fatigue failure and corrosion being the major problems.

14.2.4 Steelwork

Steelwork rarely causes many problems – except under conductor failures in blizzard conditions. Although rust will always be a problem, the main cause for ending the life of a steel cross-arm will be re-conductoring with a heavier conductor due to capacity problems. The support tie-bars are likely to fail earlier than the cross-arm due to their flexibility and less substantial construction.

14.2.5 Poles

Wood poles can have a lifetime of up to 90 years and are probably one of the most reliable parts of the network. However, the vast spread of pole strengths, even when new, can make the actual pole strength very uncertain. As the poles get older the spread of strength values widens. In some cases, therefore, an apparently sound medium pole may soon have the actual strength of a light pole. This may explain why so many sound poles fail under blizzard conditions.

Poles can be inspected by a variety of methods (see chapters 12 and 15) and the condition assessed by many devices including ultra-sonic and drilling methods. However, there is a need for an easy, linesman-proof technique to detect rot – the main killer for wood poles. Heartwood rot can be detected but is not imminently serious. White rot, which affects the surface wood and hence is more serious, is less easy to detect. Decay mechanisms in wood poles are the subject of the next chapter (chapter 15) in this book.

Defective poles often get cleared out in blizzards or hurricanes. One factor, which can distort historical lifetimes, was the tightening up in specification and inspection in 1959. Poles prior to this date have a looser specification and may not have been fully protected.

For the future, composite poles, which can have a well-defined strength, may require different inspection/maintenance techniques.

14.2.6 Distribution transformers

Under a consistent load pole-mounted transformers can last 40–50 years with virtually no maintenance. The only parts that may suffer would be rusty tanks and eroded arc gap electrodes. Lightning strokes can induce faults that lead to premature failure by acetylene build up or paper insulation degradation. However, apart from sudden failure due to lightning, the major cause of reduced lifetime is overloading – leading to insulation deterioration and overheating. Ferroresonance may also reduce transformer lifetimes. In the past most transformers had high loads in the winter when low temperatures reduced overheating. However, increasing summer loads due to e.g. air conditioning can lead to transformers being overloaded. Thermal capacity is a limiting factor in these cases.

Inspection can be visual or by oil analysis, testing for gas-in-oil, dielectric strength, moisture content, acidity and fufuraldehyde.

The temperature at which normal ageing of the winding's paper insulation occurs is 98 °C. The life of insulation, as affected by deterioration due to temperature, is given by the Arrhenius law. In the range of 80 to 140 °C, the Montsinger relation can express the law that every 6 °C deviation above or below 98 °C either doubles or halves the rate of use of life. As most transformers are operating well below 98 °C, the ageing of the electrical insulation is very slow and a forty-year life is normal. Short-term operation above 98 °C has little effect on overall life, e.g. two weeks at 110 °C would be equivalent to two months' use of life, i.e. four times normal usage.

For large transformers the limits of current, hot-spot temperature and temperature of metallic parts in contact with insulating material are lower than for smaller transformers because of the more significant stray leakage flux which causes additional heating. An additional reason for more conservative limits for large transformers is that the consequences of failure are more severe.

The position of the tap-changer either inside or outside the main tank will affect the operating temperature of the tap-changer and therefore its deterioration. The contact resistance of the tap-changer could increase at elevated temperature in line with those of other switchgear.

14.2.7 OHL joints and terminations

Two main factors contribute to failure of power connectors, namely the temperature in the interface of the connector and the temperature of the connector bulk. Essentially, the interface temperature depends upon the voltage across the connector and the bulk temperature. The failure condition is considered to be when the bulk temperature causes thermal damage to its surroundings or when the interface temperature exceeds 120 °C. The bulk temperature is determined by the I^2R losses and the thermal dissipation of the component. The application to overhead and underground systems is considered separately in the following sections.

14.2.8 Protection

Lightning protection on overhead lines is based on arc gaps or surge arresters. Arc gaps deteriorate by arc erosion in lightning storms and by rusting in salt environments. Surge arresters can deteriorate by leakage through end seals, damage to the surface by tracking, gradual performance decay by operation at high current levels in storms and finally they can suffer catastrophic failure under severe lightning overload conditions. Under ferroresonance, ungapped metal oxide arresters can fail due to long-term overloading. However, although gapped arresters are often considered as being free from ferroresonance problems, this is not always the case and experiments have demonstrated that gapped SiC arresters can fail under these conditions.

14.3 Adverse weather

14.3.1 General

The UK suffers a variable but generally mild climate. This gives the UK electricity companies particular problems that a central continental climate would not. Severe weather conditions generally occur on a short term, random basis rather than the predictable mid-continental mass periods of sub-zero conditions.

The combination of wind, snow and ice can lead to severe mechanical overloads for lines at all voltages. Small conductors (e.g. 16 or 32 mm^2 HD copper at 11 kV) can suffer excessive loads leading to strand breakage or complete conductor failure. Larger conductors (e.g. AAAC, ACSR at all voltages) can suffer extra creep due to regular overloading. This can reduce the predicted lifetime based on the initial sag and tension. Overhead line conductors can also suffer from galloping caused by small amounts of ice loading. Galloping and vibration, which can occur without ice loading, are tension dependent and can introduce strand failures due to fatigue. Reduction of these factors can allow OHL conductors to achieve their design life. In some DNOs, connections to OHL equipment or links to spurs are in small conductor sizes. These can suffer mechanical overload and fail.

14.3.2 Snow/ice

The winter season often sees night-time icing on conductors – especially on lines in upland areas. The small amounts of ice can be enough to induce vibration and galloping even on wood pole lines. The network therefore suffers stresses from these events, such as reduced conductor life, breakages near fittings and cascade pole failures.

14.3.3 Wind

The UK is open to major storms crossing the Atlantic. Winds associated with snow and ice loads can overload conductors and structures within hours if these have not been constructed to withstand these forces. Wind can also induce vibration, galloping and conductor clashing. In the past this problem has been addressed by the use of short spans and wide cross-arms with high conductor tensions.

14.3.4 Lightning

Thunderstorms can cause severe damage to OHL networks as well as being the source of many non-damage outages. Lightning damage is the major causes of damage to OHLs, both directly and also indirectly as various parts of the system (transformers, insulators, surge arresters, arc gaps) can be weakened or have their quality or efficiency reduced by lightning stress. Often this weakening of the fabric and security of the network goes undetected until some other unrelated event brings about equipment failure. Arc gaps can be eroded badly by one storm so that they fail to protect the equipment when the next one comes along. Surge arresters can suffer damage without total failure and give no obvious indication of reduced efficiency and lower protective level.

14.4 Adverse people, birds and animals

14.4.1 Vandalism

A common method of stealing conductor is to first short out the line with metal chains or similar materials, so as to trip out the circuit breaker, and then remove a length of the unenergised line. This has the effect of shortening the conductor's useful life rather dramatically.

Another aspect of vandalism, is taking pot-shots at the insulators with air rifles or shotguns, again reducing lifetime by cracking/chipping the porcelain or shattering the glass.

14.4.2 Birds

Birds can cause problems by flying into the line, taking off from equipment or cross-arms, and by nesting on parts of the line or equipment. The use of shrouds can protect hunting birds from electrocution. The occasional use of metal in nest building can short out lines near cross-arms and across bushings. Large birds flying into lines can bring the line down even when large size ACSR conductors are used.

14.4.3 Animals

Small animals such as squirrels and cats can cause damage to pole tops (and themselves) by bridging the phase wires. Rabbits can burrow into the ground at the pole base and weaken soil support. Larger animals such as cows and deer can use wood poles as rubbing posts – sometimes destroying the main strength support section in the outer layer of the pole.

14.5 Post-mortem

14.5.1 The best choice?

A line can be designed to provide power to suit the customer's needs, and it can be designed to withstand expected weather loads. However, experience has shown that a particular area may have different failure mechanisms (e.g. salt, birds, lightning levels etc.), and it may be worthwhile carrying out an analysis of what did in fact cause most economic and physical damage to the line and change the relevant policies accordingly.

14.5.2 Contracting out

These days contractors, sometimes on a turnkey basis, carry out a major part of refurbishment and new line construction. There is little time for the OHL engineer to do anything else. The problem then is to manage the design and construction so that workmanship – or lack of it – does not become a further factor in line failure.

14.6 Conclusions

Line failures can occur through gradual and predictable deterioration of line components. They can also occur through unseen vandalism or storm damage. Frequent inspection coupled with a probabilistic analysis of the benefits from maintenance can extend line life economically. However, there will always be a number of factors that no amount of economic inspection or maintenance can stop from leading to premature line failure. Some of these can be avoided not by inspection/maintenance but by policy changes. Examples of this could be improved lightning protection in susceptible areas, the use of covered conductors in forest, wildlife and leisure areas and a re-think on line design to avoid building-in future ferroresonance problems.

Chapter 15

Wood pole decay mechanisms and remedial treatments

15.1 Introduction

15.1.1 General

This chapter looks at the decay processes in wood poles used for electricity distribution networks. It covers the commonly used rot detection techniques as well as the various remedial methods employed. The use of these data to evaluate pole strength and remnant life is then discussed. The need to link wood poles to the load they are required to carry is emphasised and a simple evaluation technique is demonstrated.

15.1.2 Pole life

The serviceable life of a wood pole is determined by its capability to withstand design loads. Generally, any reduction in the strength of the pole during this period is governed by the amount of decay it may suffer due to rot. Wood poles in the UK are traditionally treated with chemicals such as creosote prior to implantation to help prevent rot and extend their service life. Other countries use a salt treatment. The long-term effectiveness of this primary treatment is determined by the extent of the penetration of the chemicals into the sapwood and heartwood, which, due to its cellular structure, does not absorb the chemicals. This chapter considers the various processes of pole decay and how (and when) it is best to deal with them.

15.2 Deterioration of wood poles

15.2.1 Wood – structure and deterioration

Wood is freely available, is naturally tough, flexible and strong and is adaptable for use. These fundamental qualities are responsible for the major use of wood for electricity distribution poles in the UK. These structures are primarily of the softwood

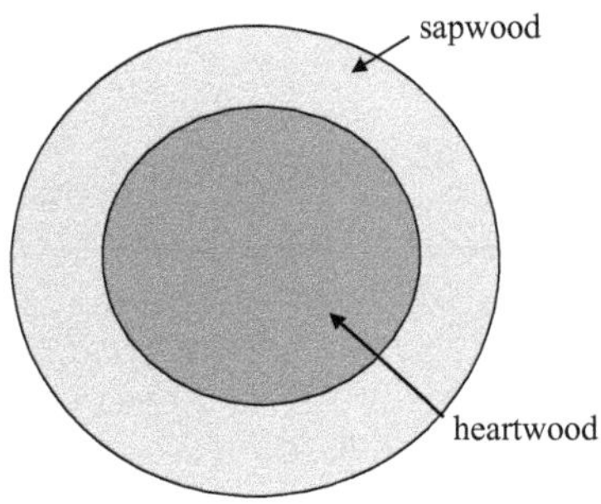

Figure 15.1 Simplified typical section of a Scots Pine pole (Pinus Sylvestris) *showing clearly defined heartwood and sapwood regions*

Scots Pine (*Pinus Sylvestris*), generally referred to as Redwood. Norway and Sitka Spruce have been tried, but their inability to fully accept preservatives applied by pressure has ensured that the poles employed in the UK are exclusively from imported Redwood stocks normally from above 60° latitude in Finland.

Softwoods possess two distinct structural zones, namely the outer sapwood and inner heartwood (Figure 15.1). The sapwood is that portion of the wood, in the living tree, that contains living cells and reserve materials such as starch, whereas the central heartwood is the dead centre that does not contain living cells or nutrient reserves [1].

Chemically, all wood structural elements consist of three main polymeric compounds: cellulose, hemi-cellulose and lignin, which form the walls of wood cells. The wood cells contain nitrogenous materials, starch, sugars and minerals [2], which form a utilisable energy source for the tree.

The wood cells also represent an attractive substrate for decay organisms, and though wood is prone to deterioration by fire, and chemical and physical decomposition [3], biological decomposition is the primary cause of deterioration [4, 5].

Decomposition by fungi represents the most sustained threat to wood poles in temperate climates such as the UK where failure due to insect attack is limited to internal timber structures. In some areas, however, woodpeckers can cause severe physical damage to the sapwood layer.

15.2.2 Pretreatment of electricity distribution poles

Wood must be moist or in conditions of high humidity for fungal decay to occur. Maintaining the moisture content of wood below 20 per cent renders it virtually immune to microbial degradation [5]. Therefore wood structures such as distribution poles, in constant contact with the predominantly moist soil conditions found in the UK, are at great risk of microbial degradation and consequent failure.

The pretreatment of electricity distribution poles in the UK is undertaken by the application of creosote oils [6] containing a high proportion of high boiling fraction oils [7] with hydrocarbons, tar acids and tar bases [1].

Prior to preservative treatment the poles must be seasoned by air-drying to a wood moisture content below 24–30 per cent, the fibre saturation point. At this point all free water within the cell spaces, the lumina, has evaporated, leaving the wood cell walls saturated with water. This water is bound chemically and physically to the cellulose, hemi-celluloses and lignin and its removal during seasoning is much slower than that of any free water. In the UK creosote is traditionally applied to the pole by the Rueping empty cell process. This process coats the wood cell walls of the treated timber with creosote leaving the interior of the wood cells almost empty of preservative.

This requires less preservative and causes less bleeding of creosote to the surrounding soil – an environmental concern. The process consists of several stages: compressed air, flooding with creosote at >65 °C and a final vacuum to remove excess creosote. Properly undertaken pressure creosoting will ensure that the entire sapwood region of a Scots Pine pole is penetrated with the preservative.

15.2.3 Problems of inadequate pretreatment

Inadequate penetration of creosote into the pole may be due to poor pretreatment seasoning, insufficient treatment pressure and/or shallow depth of sapwood.

Well-seasoned timber accepts pretreatment preservatives more freely, but preservative treatment of poorly seasoned timber is inhibited by water already in the cell lumina and by back-pressure of trapped air. Poor seasoning of timber prior to treatment was certainly a problem pre-1960 in the UK when demand for treated poles outstripped supply.

15.2.4 Pole decay and decay fungi

Fungi are the major agents of decay and of failure of wooden overhead line supports in the UK. Wood rotting fungi are classed as either soft-rot fungi [8] (*Ascomycetes* and *Fungi Imperfecti*), or brown/white rot fungi (*Basidiomycetes*).

Soft-rot fungi, as their name implies, cause a softened external pole surface. They produce cavities within the cellulose layers of wood cells by their ability to secrete cellulose. They are soil dwelling and require high moisture contents and a ready source of nitrogen to thrive in the timber. Soft rot in timber is therefore usually confined to the outer few millimetres of exposed wood (a limit imposed by oxygen supply [9] as the fungi inhabit the aerobic surface layer of soil). Soft-rot fungi therefore cause external decay around the circumference of the pole at or just below the groundline.

The *Basidiomycetes* (brown/white rot fungi) are the greatest threat to the strength and durability of exposed timber in the UK. They have no need for high moisture contents or levels of protein in excess of those already present in wood. They are not soil dwellers but proliferate via airborne spores. Brown rot fungi degrade the carbohydrate fraction of the wood whereas white rot fungi, producing cellulose and ligninases, attack carbohydrate and lignin within wood cells [5, 10]. These fungi gives rise to internal decay of distribution poles.

15.2.5 Internal and external decay of creosoted distribution poles

Decay in distribution poles is mainly found at the groundline [11–15], where moisture and oxygen conditions combine to provide an environment conducive to the growth of decay fungi. External decay leading to pole failure occurs when concentrations of pretreatment preservatives, such as creosote, begin to fall to non-toxic levels. The speed with which this reduction takes place is considered to be dependent upon the initial quantity of creosote applied [12, 16], the quality of the creosote and ageing of creosote itself [17].

Internal decay of creosoted poles occurs when the creosote treatment fails to penetrate completely the decay susceptible sapwood, due to inadequate treatment at the plant or because the wood was not properly seasoned beforehand. Poles with fully treated sapwood will still succumb to internal decay although the onset will be delayed due to the greater decay resistance of the heartwood. If pole sapwood is left untreated there is a high risk of it becoming infected with spores of wood rotting fungi which gain entry to the pole via checks and splits caused by expansion and contraction stresses in the wood during drying in service.

Untreated sapwood has virtually no natural resistance to attack by wood destroying fungi and, when infection occurs, internal decay will develop whenever the wood becomes moist enough. When decay has become established in untreated sapwood, it will eventually spread into the more resistant heartwood. In an examination of 220 000 creosoted pine poles in Germany, it was established that inadequate creosote penetration always resulted in internal decay [18].

External decay can be potentially the more serious decay problem. The outer 50–75 mm of shell in a typical distribution pole contains 80–90 per cent of its bending strength [19]. The strength of a 250 mm diameter pole is reduced by 14 per cent by only 6 mm decay of the outer shell, but a central core of 150 mm in diameter would have to be removed to have the same effect [13]. So a hollow pole can still be sufficiently strong to do its job.

Despite this, external decay is not perceived to be as great a threat to existing pole stocks as internal decay, for a number of reasons. External decay occurs only after fairly long service life. Remedial treatment of external decay is also technically simple and can be very effective because the decaying zone is easily seen, readily accessible and localised at the groundline. However, internal decay, though again found primarily at the groundline, can occur anywhere in the pole (e.g. top rot is a form of internal decay). The symptoms of internal decay such as voids and disintegrating timber are hidden from view and hammer testing is the traditional method used to sound the internal condition of the pole, although other methods have been evaluated over recent years (section 15.6). Internal decay can occur very early in the service life of a pole (even prior to pole erection) and if undetected will generally result in pole failure much earlier than external decay.

15.3 Supplementary treatment of wood poles

15.3.1 General

A number of chemicals are presently marketed as supplementary treatments for control of internal and/or external decay of pretreated wood poles. All these provide a fungi-toxic preservative effect and will, to a lesser or greater extent, slow down the decay process in wood poles. However, current legislation now restricts the disposal of the remedially treated timber at end of life.

The chemical treatments are of two basic types:

1 Those preservatives that require diffusing through the timber to be effective for treatment of internal decay (fluorides, borates and fumigants, e.g. methyl bromide).
2 Those that do not necessarily require diffusion for treatment of external decay (fluorides, borates, creosote, copper naphthenate and others).

The main precondition for the diffusion of most preservatives applied by non-pressure means is a wood moisture content of at least 30 per cent (the moisture providing the matrix through which diffusion takes place). However, this also means that leaching from the pole will ultimately lose all diffusable chemical input material.

Chemicals used for the treatment of external decay are not required to diffuse, but must be absorbed to some extent by the surface of the timber.

15.3.2 Fluoride-based preservatives (internal/external decay)

Up to 1988, the electricity supply industry attempted to control internal groundline decay of wood poles by the use of waterborne fluoride preservatives as a remedial treatment [7]. Early treatments consisted of toxic pastes of sodium fluoride and bi-fluorides applied to the pole surface in the groundline area and covered in impervious films sealed with bitumen. Bandages containing salt mixtures of sodium fluoride, bi-fluorides, dichromates, dinitrophenol and arsenic compounds sealed with a polythene cover superseded these treatments. Pole butts treated with these salts are now classed as special waste, due to high arsenic contents, with additional costs associated with disposal. The use of older salt mixtures (Cobra or DFA salts) was discontinued in the UK in 1986 owing to health and safety concerns over the dinitrophenol and arsenic components.

15.3.3 Boron-based preservatives (internal/external decay)

At present, boron-based preservatives are favoured for remedial treatments of poles in the UK and many other countries. Boron (or boric acid) diffuses well through both sapwood and heartwood, and has a low environmental impact [20]. Because of these properties, boron-based treatments have replaced the use of fluorides in the UK and many other countries.

The boron-based remedial preservatives registered and available for use in the UK are:

- Fused rods (control of internal decay): solid di-sodium octoborate for insertion into predrilled holes (100 per cent active ingredient).
- Pastes/gels (control of internal decay): boron pastes consisting of di-sodium octoborate in a glycol carrier (40–50 per cent active ingredient) for pump application through predrilled holes into internal decay voids.

- Liquids (control internal/external decay): boron liquids consisting of di-sodium octoborate in a glycol carrier (20 per cent active ingredient) for pump application through predrilled holes into decay voids (internal decay) or into plastic wraps secured around the groundline of the pole (external decay).
- Dry forms (control of external decay): boron pills affixed to the interior of plastic wraps secured around the groundline of the pole.

Unlike fluoride preservatives, boron-based treatments do not lose efficacy via gaseous loss from the pole but they are prone to leaching with subsequent reduction and loss of preservative effect.

15.3.4 Creosote (external decay)

To meet current legislation, recent developments in creosote treatment in the UK have examined the potential for a groundline wrap system which contains creosote sandwiched between layers of polypropylene (PP) and polyethylene (PE). PP forms the external layer and PE forms the internal layer against the pole surface. The wrap is applied to the pole surface and the PE internal layer is degraded by the creosote that then permeates the pole surface. There is no contact between the operator and the creosote using this system. Such a system is based on traditional methods whereby creosote emulsion bandages applied to the pole groundline successfully replenish creosote loadings in the surface of the pole such that external decay would be prevented or stopped for periods of between 10 and 15 years [13, 14].

15.3.5 Recommended control of internal decay

The treatment chemicals recommended for control of internal decay in pretreated poles in the UK are those based on boron (rods, pastes and liquids). Boron-based chemicals are the treatments of choice because:

- the treatments work and indications are that although leaching from the timber does occur, fungi-toxic concentrations are maintained for longer periods than previously thought
- there is no loss of preservative effect due to gaseous loss of preservative (as is the case with fluorides and fumigants)
- boron preservatives are freely available in the UK
- there are no HSE problems regarding disposal of timber treated with borates as their environmental profile is generally excellent
- borates are inherently safe with very low human/animal toxicity – there are therefore only minor health and safety issues regarding worker/preservative contact
- borates can be utilised in forms such that worker/preservative contact does not occur.

15.3.6 Control of external decay

Creosote (in the form of a groundline wrap) is used for the control of external decay in pretreated poles in the UK, for the following reasons:

- groundline applications of creosote in bandage form have been shown to alleviate the symptoms of external decay for 10–15 years

- creosote wrap systems designed to negate worker/preservative contact are likely to be on the market very soon, therefore health and safety problems can be reduced
- as creosote, in this situation, is being used to supplement an already creosoted product, additional end of life timber disposal problems related to the supplementary treatment cannot arise
- application of copper naphthenate would also have a beneficial effect against external decay but the copper content of this preservative may give rise to future additional end of life disposal costs (based on EU directives regarding heavy metal contaminants)
- application of borates and fluorides would also have a beneficial effect against external decay but external applications are high-leach situations and so beneficial effects are likely to be short-lived.

15.4 Application of additional preservative

15.4.1 Suitability of specific boron-based inputs for treatment of internal decay

Several utility companies have used solid rods of di-sodium octoborate periodically, principally as a preventative or prophylactic treatment. The pole is predrilled, the rods inserted and the drill holes sealed. For most practical purposes approximately nine drill holes are made into the structure to accommodate the rods and ensure a toxic effect throughout the groundline region. Boron pastes are for remedial use and are pumped into any discovered internal decay voids to sterilise the immediate area. Frequently the two treatments are used in combination with the rods inserted around the treated decay void to prevent further spread of the decay organisms.

The effects on pole strength of the number of drill holes required for rod insertion has always raised concerns and recent unpublished findings indicate that pole strength can be reduced by up to ten per cent by rod insertion. Furthermore, renewal of the rods at the end of useful life requires time-consuming field operations to crush the remaining rod remnants to the base of the drill hole or extract them. Although there are no comparable renewal difficulties associated with the use of boron pastes, they do require that significant decay must already have occurred to provide the void to accommodate the paste. Leaching of preservative is encouraged in these circumstances as the void and the decaying timber surrounding it generally behave as an effective moisture sink. However, excessive preservative usage may have some environmental and cost implications.

The most significant difference between solid rods and glycol-based pastes, however, is their differing diffusive capabilities. This is extremely important in terms of fungicidal efficacy and can be summarised as follows:

1. Glycol/paste-based boron preservatives diffuse much faster than rods at lower moisture contents, i.e. moisture contents below 30 per cent.
2. Rods generally require a minimum of 30 per cent moisture content for effective diffusion to occur. At moisture contents <30 per cent where rods do not diffuse, decay can still occur.

3 A study of 240 service poles in a high rainfall area of the east of Scotland indicated a mean groundline moisture content of 25.61 per cent [21–23].
4 Glycol preservative (paste/gel) diffusion is more even than for rods and avoids preservative spikes (i.e. areas where there is no preservative intermixed with areas where there are excessive amounts).

Therefore:

1 The prophylactic use of preservative to preempt decay (rods) represents a better approach to the problem of internal decay than remedial preservative use after decay has begun (paste).
2 The paste encourages significant leaching loss of preservative.
3 Boron pastes/gels display more efficient diffusion and are therefore more effective than rods.
4 So preventative use of pastes/gels represents the best option and is now available.

15.4.2 Supplementary creosote wraps for treatment of external decay

As indicated in section 15.3, a creosote wrap system can be used for control of external decay in standing poles, as it is a proven technique with no disposal problems. The wrap is applied to the groundline of the pole after excavation to an approximate depth of 35 cm. The wrap is then stapled in place and the excavation back-filled ensuring close contact between the pole surface and the wrap. Over a period of months, the creosote in the wrap and any remaining creosote on the pole surface will degrade the internal polyethylene layer. The creosote from the wrap will then permeate the pole surface thereby renewing the creosote loading in these areas.

Normally, such a system is re-applied in 10–15 year cycles to standing poles over 30 years old. Over such timescales the wrap will be completely denuded of creosote and no wrap disposal problems should arise.

15.4.3 Available devices for supplementary application of boron pastes/gels for treatment of internal decay

An example of a prophylactic device is shown in operation in Figure 15.2 [24]. The device introduces a measured charge of boron paste/gel into the groundline of a utility pole from an external reservoir over a period of three to four years. This device is constructed of creosote resistant engineering plastic and consists of a nozzle (with central delivery hole), a preservative reservoir (for insertion into the nozzle) and a compressed steel spring assembly and piston (to propel the preservative from the reservoir through the nozzle and into the pole). Once in position the whole assembly is made secure with a robust re-usable polypropylene cover attached to the pole. The reservoir/piston assembly tube is approximately 220 mm long with a diameter of 45 mm.

The unit is attached to the pole via a horizontal drill hole (13 mm bore) positioned approximately 150 mm above the pole groundline. It is intended to remain in position for the entire service life of the pole.

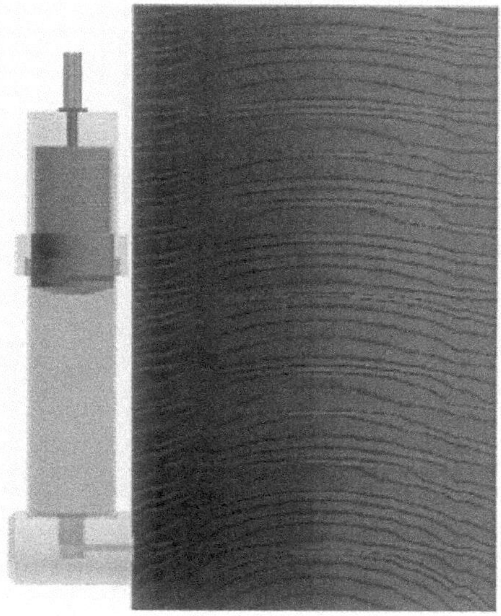

Figure 15.2 Liquid boron reservoir (courtesy EA Technology Ltd)

Approximately five minutes is required for the initial fitting of the assembly and positioning of the cover. The manufacturers recommend that initial fitting is undertaken ten years after pole erection to preempt incipient internal decay and that further preservative reservoirs be fitted to the pole at ten-year intervals.

15.4.4 Field liner primary treatment (external decay)

The field liner (FL) is a sleeve, the most recent upgrade of which consists of a robust outer layer of polypropylene woven geotextile with attached polypropylene straps over an attached inner layer of polypropylene. The inner layer of PP contains an environmentally friendly dry film biocide to prevent the growth of creosote detoxifying mould fungi at the pole/sleeve interface. All materials used are creosote resistant. The FL is simply hand fitted to the entire base of the creosoted pole (from the butt to the entire decay susceptible groundline region) prior to erection. This operation requires approximately five to ten minutes. A small diameter aperture is maintained in the FL at the base of the pole butt to prevent water logging and the development of anaerobic decay processes.

The FL restricts the occurrence of external decay in several ways:

- It severely restricts pole invasion by soil dwelling soft-rot fungi, and bacteria and non-decay fungi that detoxify the creosote preservative and provide the added nitrogen necessary to allow soft-rot fungi to proliferate in the timber.

Figure 15.3 Field liner being fitted onto new poles (courtesy Bio-Trans)

- It also significantly reduces creosote movement from the susceptible groundline area into the surrounding soil. This results in a gravitational increase of pretreatment preservative loadings in the groundline region resulting in a significant improvement in long-term preservative efficacy and a reduction in preservative contamination of surrounding soil.
- The gravitational flow of creosote to the pole butt provides enough pressure to increase the penetration depth of creosote in the groundline region.

The simple construction of the FL permits easy access for inspection of the groundline area. The application of supporting timber baulks to the pole below the groundline will not be prevented by the presence of the FL or lessen the efficiency of the system as these are generally attached below that depth at which external decay-causing fungi are active.

An alternative currently used in Australia and South Africa is the pole liner (Figure 15.3) that seals the base section of new poles before installation.

15.4.5 Top rot

Service life improvements will only be achieved if both treatments for internal and external decay are adopted. Although these devices will significantly increase the service life of utility poles, maintenance of the pole groundline over these very extended periods will result in currently minor pole problems, such as top rot in the UK, assuming much greater importance in the future. At present, very few poles fail due to this pattern of decay and if this occurs it is usually due to a failure at the preservative pretreatment stage prior to pole installation. However, the adoption of groundline protection devices will significantly extend the weathering period of the above ground part of the pole leading to significant loss of preservative efficacy here.

Figure 15.4 Installing mechanical wood pole supports (courtesy Pole Foundations)

It is likely therefore that currently rare and minor conditions such as top rot will occur with more significant effects within the service life of all poles.

15.4.6 A mechanical alternative

A pole that has suffered severe groundline rot can be fitted with a mechanical support (Figure 15.4). These can be used in urban or roadside environments where machine access is available. Such supports have already been trialled in a UK DNO.

15.5 Pole maintenance strategies

15.5.1 Introduction

The distribution overhead line wood pole population could be considered as, relative to age, reaching a critical peak. Poles introduced prior to the early 1960s are at greatest risk as it was only in the early sixties that improved standards of creosoting were introduced to the UK. The electricity supply industry introduced a considerable amount of new overhead line network in the 1950s with its rural electrification programme. There are therefore considerable lengths of overhead line in the UK with wood poles from this period.

Utilities generally perceive that wood poles have an average life of around 40–60 years, and so a significant percentage of the network requires upgrading now.

Current policy on condition monitoring ranges from any pole showing rot to be removed, to poles with a measured residual strength value or reduced factor of safety limit being selected for replacement. These policies have had a significant impact on the costs associated with refurbishment programmes as the interpretation of limited rot can range considerably depending on the inspector and devices used.

The current costs associated with pole replacement are considerable. Depending on site conditions, pole usage, design requirements, labour and material the costs associated with a single pole replacement may vary from £450 to £1000 per installation (2004 costs).

15.5.2 Pole inspection

Pole inspections have generally consisted of a visual inspection followed by a hammer test. Should any doubt be expressed by the inspector that the pole is anything other than sound, then an intrusive inspection can be carried out using a Mattson incremental borer, drill or ultra-sonic device. Fibres from the pole are inspected, following the bore, and the degree of rot is determined by carrying out several bores. This part of the inspection process requires some skill, as decayed borings tend to disintegrate on removal.

This process has, in some cases, been found to be unreliable, particularly in recent years with the loss of experienced linesmen and inspectors. Inspections could fall into two distinct areas:

1 Any rot identified during an inspection results in a pole change. This is ultimately a very expensive policy, yet gives the asset owner an assurance that no rotten poles exist on the system.
2 Poles are incorrectly inspected and guesswork results in poles remaining on the network that are unsafe.

A more refined method of inspection is therefore necessary so that estimation of pole remnant life can be consistent and reliable.

The following advantages may be achieved with a reliable supplementary inspection device:

1 Clearly identifying and using a residual strength value within a data collection system will provide a much more efficient and effective pole management system.
2 The use of electronic equipment reduces the risks associated with human error during inspection. The Health and Safety Executive (HSE), for example, has indicated that it is in favour of adopting, 'technologically proven equipment to inspect wood poles to remove the risk of mis-interpretation of results based solely on his/her experience and training'.

If the residual strength of poles can be accurately determined in a consistently reliable way then poles can be selected that have excessive decay and removed from service. Poles that remain can then be treated with a remedial treatment to attack and eliminate limited decay. This can also provide a protective barrier for those poles that have not, as yet, been infected, so ensuring that the maximum service life is achieved.

Residual strength evaluation should be compared with the load the pole is required to take with the appropriate safety factor.

A number of instruments have been introduced in recent years to address this particular issue, such as:

- sonic/ultra-sonic devices (stress and velocity)
- x-ray and nuclear magnetic resonance
- decay detecting drill
- electrical resistance instruments.

Various countries have adopted a number of acceptable residual strength values. Sometimes these are simply static figures e.g. 80 per cent of the initial strength. This can be an economic waste if the pole only requires 20 per cent of its original strength to hold the line up. Poles introduced under some wood pole design specifications have had factors of safety of 3.5, this could suggest that residual strength limits of 80 per cent are well within the pole's ultimate strength tolerances, based on BS 1990. Research is available, therefore, that could be developed to reduce the acceptable residual strength levels still further, and yet provide acceptable margins for safety and operational service.

15.6 Implications of successful remedial treatment programmes

There are a number of issues and implications that can be derived from the information provided in this chapter:

1 The effect of a lower predicted rate of decay, and hence an extension of the average useable life of a distribution/transmission overhead line wood pole through remedial treatment, will reduce network investment costs.
2 Using residual strength values within a data collection system will provide a much more efficient and effective pole management system. This will allow planners, in the future, to predict project costs and target appropriate circuits for maintenance and/or re-build.
3 A remedial pole treatment policy applied now can have short, medium and long-term implications for the wood pole population.

Based on sound research and field tests the reduction in decay rates by the use of effective remedial treatment can be as given in Table 15.1.

The 65 and 50 per cent values represent conservative technical predictions based on work with secondary preservative treatments. Pole replacements can therefore be reduced dramatically providing significant savings.

15.7 Wood pole evaluation process

15.7.1 General

Some countries rely on a purely visual inspection technique for use by foot or helicopter patrols. The inspectors are provided with a selection of photographs on which

Table 15.1 Predicted reduction in decay against pole age

Pole age (years)	Reduction of decay rate (%)
1–20	90
20–60	65
>60	50

to compare the pole condition and thereby classify it into one of (normally) five categories. These data are then fed into a condition-based decision-making programme.

When inspecting a wood pole it is important to understand the existing wood pole's design requirements based on its geographical location, the conductor it is supporting and therefore the loadings applied. Structures that are at an angle or terminal position will have additional load applied to the wood pole through the introduction of stays.

15.7.2 Strength determination

The determination of the reduced strength of the wood pole at any location is relative to the rot area and its diametrical position to the cross-section of the pole. A residual strength value can then be derived from a basic calculation or the information entered into a specific computer software program.

The decay information can be determined using a Sibert drill (microprobe high speed 7000 r.p.m., 1 mm diameter drill) process (Figure 15.5), the results of which can be converted into a graphical representation. The decay information can also be identified visually or by Mattson bore techniques (Figure 15.6) and entered manually by the operator drawing the area of rot identified.

Determination of the wood pole's capability to withstand design loads can be calculated on site relative to:

- conductor (cable) type
- conductor number
- wind span
- angle of deviation
- number of stays
- stay spreads
- pole grade
- pole height
- pole burying depth
- reduced per cent RSV.

The above parameters are used to determine the maximum permissible wind span loading for an unstayed structure and also the maximum permissible strut loading for

Figure 15.5 Sibert high-speed drill (courtesy Poletecs Ltd)

Figure 15.6 Mattson borer (courtesy Poletecs Ltd)

stayed or unstayed structures as appropriate. All of the information required to make the calculations is either referenced from the computer software or observed on site. This information is immediately used to generate the appropriate guidance necessary for the inspector to accept or reject the wood pole.

The strut loading and windspan calculations are based on loaded design conditions and factors of safety. If the revised pole strength value falls below the maximum permissible strut loading or windspan value specified the pole will fail. If the onsite design loadings are acceptable with the reduced pole strength capability then the pole will pass. This is subject to the rot identified being treatable with a remedial treatment process to slow or eliminate continued deterioration.

An estimation of residual life can be considered through analysing the information provided and assessing the wood pole's ability to be remedially treated effectively.

15.7.3 Residual strength calculations

A heartwood radius R_i (the internal radius of the sapwood shell) can be established from drilling and/or ultra-sonic data for the pole in question. The sapwood will have an external radius equal to the pole radius less the depth, r, of any surface rot. The strength of the pole depends on the fibre strength and the volume of the wood that contributes to that strength. If the assumption is made that the fibre strength is unchanged from new, then the relative strength of the pole compared to its new condition will be the ratio of the volumes of sound wood, i.e. the ratios of the $(\text{radii})^3$. A more conservative approach may be to assume a 10 per cent drop in fibre strength as well.

If the original pole had a radius R_o and now has a radius R_n, then the effective outer radius of the sound sapwood will be:

$$R_s = R_n - r$$

So the relative strength, S_r, will be:

$$S_r = (R_s^4 - R_i^4)/R_s \cdot R_o^3$$

R_n is obtained by running a tape measure around the circumference of the pole and r by using a metal prod manually to feel the depth of the surface rot. The calculations are a little more complex when dealing with an asymmetrical sapwood shell. The above calculations will give a percentage reduction in strength of the pole since new. Then it is necessary to use the current load on the pole to see whether it is still fit for purpose. This can be obtained by software such as that contained within ENATS 43-40 Issue 2/43-121.

15.7.4 Basic sample calculation

BS 1990 gives an expression for the ultimate load, F, in Newtons in terms of the pole radius at the critical point (normally the groundline):

$$F = \frac{\pi}{4} f \cdot \frac{r^3}{l}$$

where f is the fibre strength (this is taken as 53.8 N/mm^2) and l is the distance from the critical point to the point of application of the load (the pole top). So l is essentially the pole height above ground.

The pole markings in the UK indicate the date the pole was pretreated. In this example, say the date was 1974 – so the pole is 30 years old. The main strength of a pole comes from the annular sapwood ring rather than the central heartwood. So when the depth of external decay t_e (loss of sapwood) is measured, there is also a need to measure the internal decay. This enables the thickness of the annular shell (between the internal and external decayed areas) to be determined. If surface rot only is present, then the strength will vary as r^3 as above where:

$$r = (C2/\pi) - t_e$$

where C is the pole circumference at the groundline. If internal decay is also present then the strength varies as:

$$1 - (1 - t/r_c)^4$$

where r_c is the original pole radius and t is the thickness of sound wood left in the annular ring. A constant value of fibre strength with age may be assumed, although field data indicate a gradual deterioration in practice. The process for typical poles is then:

1 Calculate the ultimate load, F, for the pole that is buried a determined distance into the ground.
2 A specific depth of external rot is found on the pole. The relative loss in strength of the pole can be calculated now.
3 On another pole there is no external rot but by drilling the heartwood core is found to have severely decayed, leaving a central void 100 mm in diameter. The relative loss in strength of the pole can be calculated from this information now.
4 A third pole has lost 100 mm diameter of heartwood and has 10 mm depth of external decay. The relative loss in strength of this pole can be calculated.
5 For the third pole the rates of decay may be assumed as:

internal rot	2.5 mm per year on radius
external rot	0.5 mm per year on radius

The calculations should give the actual percentage residual strength left and this can be compared with the criterion for pole removal (possibly 65 per cent) to determine whether it should be passed.

15.8 The PURL ultra-sonic tester

15.8.1 Computer analysis of PURL results

The pole ultra-sonic rot locator (PURL) is made by EA Technology. It is operated at several points along a circumferential line around the pole and the readings obtained are then analysed using a computer program, which produces an estimate of the likely areas of internal decay and predicts the residual bending strength of the pole at the chosen cross-section (expressed as a percentage of the original strength). A decision on whether to replace the pole or allow it to remain in service is made based upon this analysis. It is therefore important that the program is able correctly to interpret the results to minimise the risk of unacceptable poles not being replaced.

15.8.2 PURL inspection process

The PURL inspection will consist of three defined stages:

1 Where the requirement for an inspection is specified then the PURL instrument will be applied at the groundline. Should, during the hammer test, there be an indication of possible decay within the pole then the PURL instrument should initially be applied at that point on the pole and then at the ground line.

The results of the initial scan are assessed onsite and should the residual strength value be below the required value, e.g. 80 per cent, then the resultant value shall be recorded and the pole will be identified with a D label. If the residual strength value is greater than 80 per cent then stage 2 will be initiated.

2 As stated in stage 1 this procedure shall be initiated if the ground line test has resulted in a residual strength recording of greater than 80 per cent. The PURL instrument will be applied at approximately 500 mm above the ground line and a test at this position on the pole shall be carried out.

The results of the initial scan will be assessed onsite. Should the residual strength value be below the required value, e.g. 80 per cent, then this value will be recorded and the pole will be identified with a D Label. If the residual strength value is greater than 80 per cent then stage 3 will be initiated.

3 Where inspection of the pole is necessary below the ground line, the extent of such excavation shall not exceed 300 mm below the ground surface. The circumference of the pole will be exposed and the PURL inspection will be carried out below the ground line area and the results recorded. The appropriate A or D label shall then be fixed to the pole relative to the results found.

On completion of the test the area shall be backfilled and returned to its original condition; moderate compaction of the backfill should be achieved.

15.8.3 Residual strength value (RSV)

The generally acceptable RSV in the UK is 80 per cent of its original condition based on values from BS 1990 Part 1: 1984. Any poles falling below this residual strength figure have to be condemned and therefore require replacement.

If it is not possible to ascertain the actual RSV figure during the inspection process (onsite) all poles suspected of having decay have to be labelled appropriately with a D label until such time as the investigative data can be scrutinised. Current regulations dictate the actions in Table 15.2.

Table 15.2 Actions based on RSV

Urgent	poles that are inspected and are found to be a threat to public safety due to the internal decay or external damage should be reported and dealt with immediately; residual strength value (RSV) of less than 50% (D label)
Defect	poles that are found to have a residual strength value (RSV) of less than 80% will be identified appropriately for future removal (D label); poles identified to have considerable evidence of soft rot shall be identified appropriately for future removal (D label)

15.9 Cost benefits

Whatever the method of testing poles, the advantage of going beyond the hammer is a cost-effective pole management programme. Typically, the following results were obtained in field tests on 11 kV poles (courtesy Poletecs Ltd):

- 948 poles identified by hammer as being suspect (S) or decayed (D)
- re-tested using ultra-sonic tests
- 525 of the re-tested poles (55.3 per cent) re-classified as sound poles (A)
- 277 of the re-tested poles (29.2 per cent) re-classified as having some rot (D2)
- 146 poles (15.4 per cent) confirmed as needing to be changed (D1).

to replace the 948 poles at, say, £600 a pole, would be		£568k
the cost of testing at £50/pole would be	£47.4k	
to replace 146 rotten poles would be	£87.6k	
to treat 277 poles at £50/pole would be	£13.85k	
Totals	£148.85k	£568k

That equates to a saving of around £420k or £443 per pole.

Another way of estimating saving is to cost the saving of delaying pole expenditure by ten years. At five per cent per annum interest saved on £600, this amounts to a saving per pole of £324.

15.10 References

1 WILKINSON, J. G.: 'Industrial timber preservation' (Associated Business Press, London, 1979)

2 KIRK, T. K.: 'The chemistry and biochemistry of decay' *in* NICHOLAS, D. D. (Ed.): 'Wood deterioration and its prevention by preservative treatments. Vol. 1. Degradation and protection of wood' (Syracuse University Press, New York, USA, 1973), pp. 149–182

3 LEVY, J. F., and DICKINSON, D. J.: *in* ROSE, A. H. (Ed.): 'Economic microbiology. Vol. 6. Microbial deterioration' (1981) pp. 19–60

4 CARTWRIGHT, K. St. G., and FINDLAY, W. P. K.: 'Decay of timber and its prevention' (HMSO., London, 1958)
5 SCHEFFER, T. C.: 'Microbial degradation and the causal organisms' *in* NICHOLAS, D. D. (Ed.): 'Wood deterioration and its prevention by preservative treatments. Vol. 1' (Syracuse University Press, New York, USA, 1973)
6 BS 1990: 'Wood poles for overhead power and telecommunication lines. Part 1. Specification for softwood poles', 1984
7 BS 144: 'Specification for coal tar creosote for wood preservation', 1997
8 SAVORY, J. G.: 'Breakdown of timber by ascomycetes and fungi imperfecti ann', *Appl. Biol.*, 1954, **41**(2), pp. 236–247
9 SAVORY, J. G.: 'The role of micro-fungi in the decomposition of wood', *Rec. Ann. Conv. B. W. P. A.*, 1955, **5**, pp. 3–19
10 LIESE, W.: 'The action of fungi and bacteria during wood deterioration', *Ann. Conv. B. W. P. A.*, 1970, **4**, pp. 81–96
11 CHAMBERS, L. G.: 'In-situ treatment of poles – part III, groundline treatment using preservative salts', *Aust. Telecommun. Monogr.*, 1963, (2), pp. 103–105
12 SMITH, D. N., and COCKCROFT, R.: 'The remedial treatment of telephone and electric transmission poles. Part 3. Treatment for internal decay', *Wood*, 1967, **32**(11), pp. 29–31
13 SMITH, D. N., and COCKCROFT, R.: 'The remedial treatment of telephone and electric transmission poles. Part 1. Treatment for external decay', *Wood*, 1967, **32**(9), pp. 35–37
14 SMITH, D. N., and COCKCROFT, R.: 'The remedial treatment of telephone and electric transmission poles. Part 2. Treatment for internal decay', *Wood*, 1967, **32**(10), pp. 37–40
15 BECKER, G.: 'Treatment of wood by diffusion of salts', *J. Inst. Wood Sci.*, 1976, **7**(4), pp. 30–36
16 SMITH, D. N.: 'Field trials on coal tar creosote and copper/chrome/arsenic preservatives: results from a new method of assessment', *Holzforschung*, 1969, **23**, pp. 185–192
17 DICKINSON, D. J., MCCORMACK, P. W., and CALVER, B.: 'Incidence of soft rot in creosoted wooden transmission poles'. The Inter. Res. Group on Wood Preserve. Document IRG/WP/1554-92, 1992
18 WARRELMANN, E.: 'Findings about creosote and salt impregnated timber poles on the basis of the statistics of a large electric power board', *Elektrizitatswirtschaft*, 1956, **55**(23), pp. 869–875
19 BINGEL, N. G.: 'Wood pole restoration'. Proceedings of Wood pole conference, (MORRELL, J. J. (Ed.)) Portland, Oregon, USA, 1988, pp. 81–89
20 SINCLAIR, D. C. R., SWEET, R., BRUCE, A., DURRANT, P., and SMITH, G.: 'Assessment of the efficacy of a remedial treatment for creosoted utility poles'. Proceedings of the international conference on *Utility line structures*, Fort Collins, Colorado, USA, 1998
21 SINCLAIR, D.: SIWT Projects Ltd, private communication

22 EDLUND, HENNINGSON, KAARIK and DICKER: 'A chemical and mycological evaluation of fused borate rods and a borate/glycol solution for remedial treatment of window joinery', 1983
23 AMOFA: 'Penetrability of boracol solutions in Scots pine sapwood (*P. Sylvestris*) of varying moisture contents and diffusion periods'. Wood and Cellulosics. ISBN 07458-0113-7, 1987
24 HORSMAN, D., and SINCLAIR, D.: 'Accumulation of data relating to the decay and remedial treatment of wood poles', EA Technology report 5416, April 2001

15.11 Further reading

BECKER, G.: 'Fluorine compounds for wood preservation', *J. Inst. Wood Sci.*, 1973, **6**, pp. 51–62

Chapter 16

Line and component susceptibility to weather effects

16.1 Scope

16.1.1 Components

The various components of an OHL that will be considered are:

- conductors
- insulators
- poles
- supports and fittings
- foundations
- pole-mounted transformers
- pole-mounted switchgear
- lightning protection – arc gaps
- lightning protection – surge arresters
- line design.

16.1.2 Weather environment

The weather effects that will be considered are:

- lightning
- snow
- ice
- wind – gales
- wind – low wind speeds
- pollution
- temperature
- others.

Lightning is covered in detail in chapter 18; however, the effects on particular components are also covered briefly here for the sake of completeness.

16.1.3 Procedure

Each OHL component will be considered for all relevant weather effects. There will inevitably be an interaction with other components. These will be considered at the time and backreferenced later.

16.2 Conductors

16.2.1 Lightning

Lightning can affect overhead lines, either by:

- direct strike, causing the lightning strike current to travel along the conductors until a route to earth is found
- an indirect strike where a voltage surge is induced in the line by the magnetic field associated with a strike to an objector ground near the line.

16.2.1.1 Bare conductor

Direct strike

Neither a direct nor indirect strike to a line will cause conductor damage directly, except in extremely rare occasions in the case of a direct strike with a high strike current (>100 kA), where localised heating at the struck point may be sufficient to cause strand failure. This has sufficiently low probability not to be considered as part of a condition assessment.

Local arcing

Lightning overvoltage surges on OH lines can cause air breakdown across insulators at low BIL points. Such points are:

a stay wires
b pole-mounted equipment
c between phases.

Items *a* and *b* can occur as a result of direct and indirect strikes. Item *c* can occur for direct strikes only. In item *a*, arc breakdown may occur across a pin, post or tension insulator and the stay insulator. In item *b*, the arc is more likely to occur across the bushing of the pole-mounted equipment. In both of the above cases, the arc will not move and will cause damage at the arc root (on a fitting or conductor or arc electrode) until circuit breaker operation.

In item *c*, the arc will be struck between two adjacent phases but will travel in the direction of the load. It will not cause an earth fault until the situations of items *a* and *b* occur. No damage to the conductor is likely due to item *c* alone.

1 11 kV systems – damage to conductors and insulators likely for direct (10 per cent of damaging) and indirect (90 per cent of damaging) strikes. Geographical areas of likely damage can be calculated from lightning strike density maps.

2 22–33 kV systems – as for 11 kV, but with reduced effect from indirect strikes (80 per cent for 22 kV, 65 per cent for 33 kV).
3 >33 kV systems – susceptible to direct strikes only. Susceptibility can be calculated from lightning strike/current density maps.

16.2.1.2 OPGW

Optical pipe ground wire (OPGW) is strung on 33 to 132 kV lines. If the OPGW is strung below the phase conductors then lightning damage is unlikely, except at joints to ground-based telecommunications.

OPGW strung as an earthwire above the phase conductors can be damaged by a direct strike if it is the single-layer type (Horse) due to poor thermal conductivity. Double-layer types (Lynx or Keziah) are unlikely to suffer damage as they have more aluminium strands that can absorb the generated heat from the arc. Damage levels can be calculated from lightning strike density maps.

16.2.1.3 Covered conductors

Covered conductors are susceptible to lightning damage due to phase–phase or phase–earth arc breakdown. This is likely to cause conductor failure in medium voltage systems. The application of appropriate lightning protection should reduce this susceptibility. Such protection (APDs or PADs) is not normally susceptible to significant deterioration over time.

16.2.1.4 Summary

Lightning affects conductors in specific discrete events. There is no condition information as such on which to base conductor lifetimes.

16.2.2 Snow/ice

16.2.2.1 General

Conductors can be subject to snow and ice as well as wind-on-ice loads. In lowland areas (<200 to 300 m land height), the weather will generally consist of specific wet snow incidents. Potential conductor loads can be determined from geographical location according to ENATS 43-40 wind/ice maps. Although these are discrete, individual events, line design will generally determine the lifetime of the system under the predicted weather conditions.

Snow/ice and wind-on-ice loads can be determined from historical data, as well as geographical location. Appropriate line design can reduce the stresses on OHL components by evaluating the predicted loads and allowing for factors of safety (FOS) or probabilistic risk assessment. Condition information can then determine the withstand strength of components to such loads. As far as conductors are concerned, the use of small conductors ($\leq$35 mm^2 copper equivalent) risks heavy rime ice loads in upland areas, as ice tends to grow more rapidly on small objects. Historical data from past storms have also shown that small conductors suffer proportionately far more failures than those >35 mm^2 copper equivalent.

16.2.2.2 Types

Snow and ice come in many forms and can affect the network in different ways.

Wet snow

Wet snow occurs as large flakes containing 15–40 per cent of water. Air temperatures are around 0.5 to 2 °C. It builds up rapidly on overhead lines and structures, causing wood pole network failures within two to three hours and tower line failures in six to eight hours under severe conditions. The accretion is fairly dense at around 850 kg/m^3 and tends to form circular envelopes of ice.

Dry snow

At temperatures below zero the liquid water content (LWC) of the flakes is less than 15 per cent and it does not tend to stick. This property is good for overhead lines as the conductors will not accrete snow loads. It is bad, however, for any equipment that could be damaged by ingress of snow as this very lack of stickiness allows the snow crystals to punctuate deep into crevices or through ventilation openings.

Rime ice

Rime ice can accrete on overhead lines at sub-zero temperatures down to −5 °C. It occurs in low cloud (and so is often called in-cloud icing) at ground level – a situation very common in the hills in winter. Particles of around 10 μm size are carried with the wind and accrete on conductors and supports. The initial accretion extends from the conductor into the direction of the wind and is similar in section to the aerofoil of a plane wing. This can cause uplift and resultant galloping of the conductor or stay wire. The conductor will exhibit a low-frequency, high-amplitude oscillation described later in this chapter under 'wind effects'. The motion causes severe impulsive forces at the supports.

In the long term rime ice can accrete heavy loads over a period of 24–48 hours. The ice density will be around 500–700 kg/m^3 and the combined wind and ice load can bring down overhead lines. At temperatures below −5 °C the ice tends to be friable and of a lower density (down to 300 kg/m^3) but can still cause excessive ice loads in gentle wind conditions.

Glaze ice

Glaze ice occurs when rain falls from a temperature zone through a 200–300 m layer of sub-zero air at ground level in what is known as a temperature inversion. The raindrops stay as water but in a super-cooled state so that any physical contact causes a rapid transition (less than one second) to ice. If the raindrops touch overhead lines then a rapid build up of heavy ice loads can occur, bringing lines down in an hour or so. Because the ice comes from clear water there is no trapped air and the accretion is clear, virtually pure ice with a density of 914 kg/m^3.

16.2.3 Wind

16.2.3.1 General

Wind can cause problems for overhead lines in several ways:

1. Gales – where wind force on conductors and poles can bring down OHLs.
2. Light winds – where Aeolian vibration can be set up in the conductors, reducing lifetime by fatigue at joints or other nodal points.
3. Moderate winds – can cause galloping on tower lines, but rarely on the shorter span wood pole lines.
4. Wind-on-ice.
5. Blown debris – loose debris that can damage or short out OHL conductors.
6. Blown-down trees – a major cause of line failure if inadequate tree-cutting programmes are employed.
7. Clashing.

16.2.3.2 Types of damage

Conductor loads

Wind in itself does not cause large increases in conductor or support loads but under the large surface areas presented by iced conductors this can become significant.

Vibration

Light winds can cause Aeolian vibration in conductors. The general way to avoid these vibrations (generally up to 100 Hz) is to have a maximum stringing tension for conductors depending on whether these are copper, aluminium, covered etc. The problem generally leads to fatigue failure at the compression fitting.

Galloping

Strong winds can cause uplift on conductors in long spans – especially under light icing conditions. This uplift can lead to galloping where low-frequency (1–3 Hz) high-amplitude (up to the sag value) oscillations occur in the conductor. This type of motion causes massive impulsive forces at the conductor terminations. These forces may be two to three times the conductor tension and can lead to stresses beyond the elastic limit or to fatigue failure. Antigalloping devices such as spacers or dampers can be used. This phenomenon affects single, twin and bundled conductors.

Debris and trees

Light debris can be blown across the phases of overhead lines causing temporary or permanent faults requiring re-closer operation. Snow laden tree branches can bend down on to wood pole lines causing phase–phase faults on bare wire lines but not on covered conductor lines (see Figure 9.17, chapter 9).

Major storms can cause trees to fall on to distribution lines. In the case of bare wire lines this can lead to conductor failure or asymmetric loads and subsequent pole failure (due to twisting of the pole). Narrow phase spacings on covered conductor lines can often aid survival of these three phases. This spreads the load and reduces the chance of conductor failure. It also puts a symmetrical in-line stress on the wood poles that they can more easily withstand than the twisting movement from bare wire lines.

16.2.4 Pollution

16.2.4.1 General

Pollution in the UK generally means salt pollution, although industrial pollution, such as cement dust from factories or rock dust from quarries, can also cause problems. These problems are mainly relevant for insulators but do affect conductors in two ways:

1 corrosion – on aluminium and steel
2 surface tracking on covered conductors.

Pollution comes in several forms. Salt pollution near coasts can corrode aluminium conductors and steel components (e.g. arc gaps). Industrial pollution can be carried hundreds of miles in weather systems and cause surface degradation on insulators and sheaths. Pollution can also travel in rain and snow or ice crystals and be deposited on overhead line components.

16.2.4.2 Salt

Within 10 km of the coast in the UK the high salt concentration in the atmosphere can lead to corrosion of any aluminium-based conductor as well as causing sheath deterioration to insulators and surge arresters.

Electrolytic cells are set up in the interstices of aluminium conductors. In the absence of oxygen the aluminium loses its protective layer and corrosion proceeds rapidly in the outer strands. The use of small aluminium conductors is therefore undesirable in these areas. Larger, multi-layered conductors perform better but greasing is essential.

Salt layers on insulating surfaces allow micro-discharges to occur which can give intermittent supply problems on porcelain and long-term surface tracking on polymeric insulators or arresters. Another form of corrosion is the loss of galvanising in ACSR conductors, leading to failure of the steel core.

These bare conductor corrosion problems lead to loss of strands and eventually elastic failure. Covered conductor sheaths can suffer tracking problems depending on the voltage stresses, e.g. the use of bare metal ties on high carbon content sheaths with porcelain insulators (see Figure 9.16, chapter 9).

16.2.4.3 Industrial pollution

Sulphates and other pollutants can travel hundred of miles on weather systems and cause sheath or surface problems on insulators. Lines next to cement works or to sand dunes can also suffer in the same way.

Snow and ice can become polluted in the process of accretion and thereby reduce insulation strength. This can lead to severe arcing on iced insulator strings.

16.2.4.4 Covered conductors

In severe salt pollution areas, the sheaths of covered conductors can be coated with a semi-conducting layer that allows tracking or micro-discharges to occur in:

- the gap between an earthed or electrically floating conductor fitting and, for example, the IPC of a covered conductor arc protection device or earthing point
- the area of high electrical stress at the end of helical ties.

The above problems are particularly enhanced by the use of high carbon content sheaths (~3 per cent carbon) used for UV protection. Severe problems can occur as shown in Figure 9.16 in chapter 9. The use of zero carbon content sheaths can significantly reduce this problem.

16.2.5 Temperature

16.2.5.1 General

Temperature can affect OHL conductors in several ways:

1. High ambient temperatures reduce the cooling effect of conductors and can allow highly (electrically) loaded conductors to exceed the design temperature.
2. High electrical loads can also lead to conductor temperatures above the design load.

The above factors can cause clearance problems as conductors can sag below regulatory clearance values. In some conductors, especially small copper conductors, sustained elevated temperatures can lead to annealing and softening of the material. This can allow conductors to stretch beyond their elastic limit, possibly infringing regulatory clearances.

16.2.5.2 The sun

The sun can affect the current rating of conductors but its main source of damage is through UV degradation of composite insulators and covered conductor sheaths. UV stabilisers are used to avoid this problem, which can lead to reduced resistivity (surface tacking) and physical cracking of the surface as it becomes brittle. A common and cheap UV stabiliser is carbon black but this can make sheaths susceptible to micro-discharges and radio frequency emissions (see § 16.3.6). Alternatives to carbon black include TiO_2 but these compounds can be more expensive and add significantly (up to 30 per cent) to the cost of covered conductors.

16.2.6 Other areas

16.2.6.1 Emissions

General

Overhead lines and sub-stations can affect the environment by the presence of electromagnetic (EMF) and radio frequency (RF) fields and acoustic noise (from corona). Earth faults on lines can also lead to local rises in ground potential around the earth connections of pole-mounted equipment or towers.

EMF

Electro-magnetic fields arise from the AC current flowing in each conductor. They can be reduced by suitable design of spacing and phase selection but in general this is not an option for existing lines. EMFs are seen as a cause of concern and strict limits are set for their levels for all areas of public access. However, the general consensus of scientific work is that EMF is a perceived but not a real problem. Currently, the EMF emissions from UK lines are significantly below UK and European allowable emission limits.

RF

Radio frequency emissions occur due to poor connections or surface tracking on tower and wood pole lines. They are generated by the micro-discharges that occur in these situations and can often be detected by interference on radio and TV. They are never seen as a danger to health but can have a high nuisance value. Strict emission levels over a wide range of frequencies are specified in current standards and line emissions can easily be measured. The measurements have to take place on both dry and wet days and preferably with the power on and off. This latter point is because there is a natural background level of RF and the source may not be the power line or sub-station.

Acoustic noise

The intense electric fields close to overhead line conductors can cause ionisation of the air, especially on damp, humid days. This ionisation can be heard quite clearly on heavily loaded lines. It has nuisance value only to the public but is an energy loss problem to the supply company. Levels can be measured and compared to allowable levels from current standards.

16.2.6.2 Rain

Rain is useful as it can clean pollution from overhead line components. However, localised flooding can cause problems to sub-stations built to supply power to developments in flood plains. With the likely increase in frequency of storms and short-term high rainfall incidents due to climate change, this may become an increasing problem in future years. Overloaded drains built in times of lower local population, less tarmac and concrete and more green fields are increasingly incapable of handling the rapid run-off of heavy rain.

Rain can help covered conductors to withstand polluted environments by washing off surface contamination and so reducing the amount of time for which damaging surface tracking will be present. Condition assessment can enable such factors to be evaluated in terms of HIs and their effect on conductor lifetime.

16.2.6.3 Wildlife

Falcons using overhead line transformers as perches, large birds taking off from cross-arms or flying into overhead lines – all these cause problems to the supply (and of course will generally kill the bird). Squirrels and cats can also short out phases as they use the lines as their own road crossings. Nature reserves, migratory flight paths and estuary crossings are potential danger areas for such incidents. Covered conductors and insulator shrouds can help reduce these incidents, resulting in better supply quality and reduction in wildlife fatalities.

Condition information can give no benefit to the effect of wildlife. The choice of conductor (e.g. obvious covered conductors) or other warning devices (balls on-line, bird diverters etc.) can reduce susceptibility to wildlife problems.

16.3 Insulators

16.3.1 Lightning

Lightning-generated overvoltage surges raise the voltage across the pin, post or string insulator, but only for a few micro-seconds. If, in this time, the surge is sufficiently intense to cause a breakdown over the insulator surface or in the air gap between conductor and spindle (or cross-arm) or any arc gap, then an arc will be generated. The arc will be stationary only if there is a local earth involved.

16.3.2 Arcs

Arcs can cause insulator problems in several ways:

- thermal shock
- localised surface heating
- surface tracking
- arc root damage (especially on string insulators)
- coating of insulator surface with vaporised metal (from conductor).

Thermal shock arises as the arc is actually a plasma of ions and electrons at temperatures of several thousand °C. Porcelain insulators can therefore be thermally stressed and crack or even shatter.

Localised heating can melt the porcelain and/or polymeric insulator. This can occur when follow-through current that tracks over the edges of the insulator sheds sustains the arc.

Surface tracking can occur due to the high electrical stresses in a polluted environment. Although the lightning will only provide the initiation process of tracking, this effectively reduces the insulator AC withstand and so allows higher surface currents under normal operating conditions.

Arcs travel through the air across surfaces and are rooted at the ends on conductive areas. This root will generally be a small area (~1 mm or so across), causing heating due to the locally high current density. This can easily elevate the temperature of the conductive area (e.g. top of a transformer bushing or the metal centre of a string insulator), causing local stresses at the metal/ceramic or glass interface, which can lead to localised melting and cracking.

An arc root on an aluminium or copper conductor at an intermediate pole-top or a transformer bushing can vaporise the material, which then coats the upper sheds of the local insulator surface.

Most of these problems occur due to the follow-through earth fault current that occurs following an initial lightning-generated arc.

16.3.3 Effects on BIL

The condition of insulators and bushings can deteriorate due to the effects mentioned. The insulators are then more susceptible to other environmental effects such as rain or pollution, leading to intermittent breakdown in some cases and possible poor supply quality. The lower BIL value allows micro-discharges across the surface or in cracks, making the line more susceptible in future storms.

Condition information from current leakage under normal operating voltages or visual inspections (e.g. blue arcing visible at night) can indicate deteriorating insulators.

16.3.4 Snow/ice

16.3.4.1 General

In general, snow and ice do not cause major problems to insulators. However, ice can often contain atmospheric dust that allows it to become conductive. Ice can also bridge insulator sheds. Tracking and discharges across iced insulators are relatively common but only cause damage if the arcs transfer to the insulator surface. Most problems are caused by discharge currents leading to intermittent SEF (sensitive earth fault) circuit breaker operation.

16.3.4.2 Line failure

Heavy snow/ice conductor loads leading to sudden tension changes (ice shedding or conductor breakage) can cause insulator damage, either by fracturing the porcelain/glass or bending the support spindle.

16.3.5 Wind

Wind is rarely a problem for insulators, except in the case of severe conductor movement causing broken ties, or debris blown across the cross-arm insulators. Vibration of conductors does not normally damage the insulator.

16.3.6 Pollution

16.3.6.1 General

Salt pollution near coasts can cause surface degradation on insulators and sheaths. Pollution can also travel in rain and snow or ice crystals and be deposited on sheds.

Salt

Within 10 km of the coast in the UK the high salt concentration in the atmosphere can lead to sheath deterioration on insulators and surge arresters. Salt layers on insulting surfaces allow micro-discharges to occur that can give intermittent supply problems on porcelain and long-term surface tracking on polymeric insulators.

Covered conductor sheaths can suffer tracking problems depending on the voltage stresses, e.g. the use of bare metal ties on high carbon content sheaths with porcelain insulators.

Industrial pollution

Sulphates and other pollutants can travel hundred of miles on weather systems and cause sheath or surface problems on insulators. Lines next to cement works or to sand dunes can also suffer in the same way. Problems have been noted specifically for silicon-type composite insulators near fertiliser plants.

Snow and ice can become polluted in the process of accretion and thereby reduce insulation strength. This can lead to severe arcing on iced insulator strings.

Pollution can be transferred thousands of miles and deposited in snowfall on exposed highland areas. This can lead to the discharges mentioned in the previous section.

16.3.6.2 Polymeric sheaths

Polymeric sheaths can exhibit surface tracking due to pollution and electric stress (Figure 16.1).

Very often, the support matrix material in a silicon insulator or surface damage in an EPDM insulator can exhibit tracking in areas of high electric stress, such as under the sheds near the insulator stem. This can be detected either visually or by increased leakage currents under normal operation.

16.3.7 Temperature

16.3.7.1 General

Temperature as such has little effect on insulators. Local high temperatures due to arcs have already been discussed.

16.3.7.2 The sun

UV problems occurred widely in early polymeric insulators. UV can degrade insulator surfaces from insulation levels of $>10^{12}$ Ωm to below 10^8 Ωm. At the latter level, the insulator surface is not sufficient to restrict damaging discharge currents under normal operation. These can occur in porcelain insulators, due to glaze damage, or in polymerics due to surface changes. Both can normally be identified visually by surface discolouration.

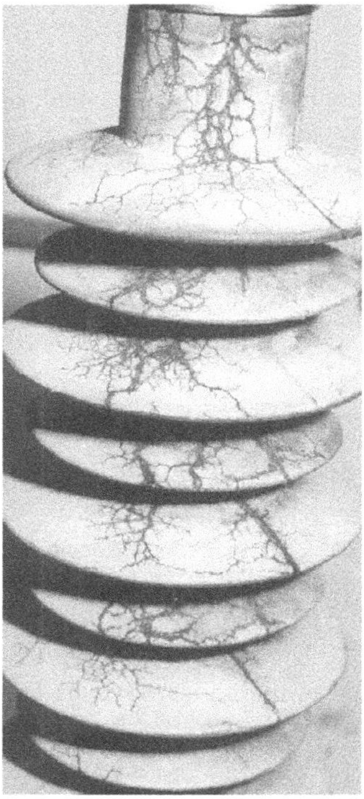

Figure 16.1 Surface tracking on insulator sheds

16.3.8 Other effects

16.3.8.1 Fog

Fog is a particular problem for insulators. It causes similar problems to pollution by the deposition of clean or pollution water droplets on insulator surfaces. In areas where sand or salt or other pollutants are already present, the presence of fog – or even overnight dew – can cause intermittent arcing and lead to tracking damage.

In susceptible areas, long creepage length insulators can be used that hide extra surface length under insulator skirts (Figure 16.2). Alternatively, material choice for insulators can reduce the flashover susceptibility, as has been demonstrated in salt fog chamber tests.

16.3.8.2 Dew and pollution

In deserts, night-time dew and sand or salt pollution can cause severe tracking on insulators and the creepage length has to be extended. If the creepage length is not sufficient then surface tracking can cause further damage, leading to shortened lifetimes and poor supply quality. Long creepage length insulators, which utilise deep corrugations on dishes, can suffer due to pollution staying within these corrugations and not being washed off in rainstorms (Figure 16.2, areas marked 'M').

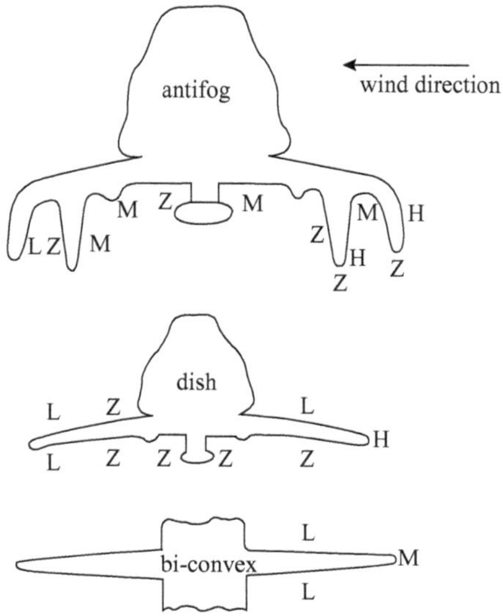

Figure 16.2 Long creepage length and alternative shed design insulators

16.4 Poles

16.4.1 Lightning

Lightning as such is rarely a problem for wood poles, although there are examples of direct strikes causing split poles due to current flow to earth.

16.4.2 Snow/ice

16.4.2.1 General

Snow/ice loads on conductors can put crippling loads onto wood poles. Appropriate line design can avoid this situation becoming a major problem by realistic load calculations based on expected weather scenarios. It has already been mentioned that small conductors ($\leq$35 mm^2 copper equivalent) can attract substantial snow/ice loads relative to their strength. These can lead to conductor failure and subsequent pole breakage due to torsional forces. Unbalanced loads can also lead to significant bending moments applied just above ground level.

16.4.2.2 CENELEC standard

European standards (BS EN 50341 and BS EN 50423) have initiated a re-think of UK line design due to the problems associated with continuing to use the semi-probabilistic standard ENATS 43-40 Issue 1. The use of a new standard based on ENATS 43-20 has been put forward as the NNA to CENELEC in BS EN 50423-3-9 (see chapters 4 and 5). The appropriate application of line design based on this NNA should form a basis for establishing line design health indices (HIs) for pole health, although the continued allowance to apply zero ice loads for small conductors is of some concern.

16.4.3 Wind

16.4.3.1 General

At distribution level, wind-on-pole evaluation has not been the norm for line design of intermediate poles. Of perhaps more significance is the wind load on conductors with wind direction normal to the line. Wind-on-ice conductor loads can cause pole or foundation failures. This situation should be resolved by line design HIs and pole strengths HIs.

16.4.3.2 Trees

A major cause of line failure is broken poles due to trees being blown onto the line by gale force winds. The likelihood of trees falling onto the line depends on an analysis of tree management and cutting techniques or right-of-way (ROW) HIs. A correctly determined ROW management scheme based on tree numbers, types, condition and distance from the line can significantly reduce the susceptibility of pole damage due to fallen trees.

16.4.4 Pollution

Pollution as such has little effect on wood poles. Salt pollution on the surface can reduce the voltage withstand of creosoted poles by 20–30 per cent, especially in wet conditions. This may have significance when there is a high level of leakage current down a pole due to, for example, cracked insulators. Another possibility for pole-top electrical stress is where stay wires are not bonded to pole-top hardware, leading to voltage gradients which could cause high damaging leakage current levels to be present.

16.4.4.1 Temperature

Temperature has an effect on moisture levels in wood poles and hence an effect on rates of fungal decay. This also has an effect on the rate of diffusion of pole treatments, such as boron rods into the pole. Low moisture levels ($<$25 per cent) can reduce boron diffusion and allow fungal decay to go on unchecked.

16.4.5 Other effects

16.4.5.1 General

The biggest threat to wood pole strength and hence the ability to hold up an overhead line lies in attack by:

- fungi
- insects
- birds.

Initial choice in the sourcing of wood poles, correct pretreatment procedures followed by correct primary treatment can significantly reduce fungal attack, but these have little effect on insect/bird damage.

16.4.5.2 Fungi

The serviceable life of a wood pole is determined by its capability to withstand design loads. Generally, any reduction in the strength of the pole during this period is governed by the amount of decay it may suffer due to rot. Wood poles are traditionally treated with chemicals such as creosote prior to implantation to help prevent rot and extend their service life. The long-term effectiveness of this primary treatment is reduced by the limited penetration of the chemicals into the heartwood which, due to its cellular structure, does not adsorb the chemicals (see chapter 15). Softwoods possess two distinct structural zones, namely the outer sapwood and inner heartwood. The sapwood is that portion of the wood, in the living tree, which contains living cells and reserve materials such as starch, whereas the heartwood is that portion which has ceased to contain living cells and where nutrient reserves have been removed or converted to heartwood substances.

The wood cells also represent an attractive substrate for decay organisms, and although wood is prone to deterioration by fire, and to chemical and physical decomposition, biological decomposition is the primary cause of deterioration.

Decomposition by fungi represents the most sustained threat to wood poles in temperate climates such as the UK where failure due to insect attack is limited to internal timber structures. In some areas, however, woodpeckers can cause severe physical damage to the sapwood layer.

16.4.5.3 Insects

Insects are rarely a problem in the UK. Insects such as termites can attack poles in tropical regions but the UK environment is not suitable for their existence.

16.4.5.4 Birds

Nesting birds can be a problem for wood poles, generally, though not exclusively, in northern parts of the UK due to the lack of suitable alternative nesting sites (trees). Western Scotland, the Western Isles and the Arcadians have few large trees and so birds that need to nest high off the ground tend to choose pole-mounted equipment, cross-arms or the poles themselves as nesting sites. In relation to wood pole damage there is a chance of pole-top fires started by metallic debris collected by birds shorting out OHL conductors phase–phase or phase–earth and igniting nesting material. This could then lead to the pole itself burning.

A more significant problem lies in the birds that nest within old trees and typically the woodpecker. The problem here is that it is not necessarily old or rotten poles that are selected by woodpeckers. It is not unknown for poles of only two to three years of age being weakened substantially by woodpecker damage.

Some utilities have established a condition-based assessment system to grade woodpecker damage according to size and number.

16.5 Supports and fittings

Stay wires are commonly neglected and not maintained at correct tensions. This can allow forces on the pole to be directed on an inclined and not vertical pole, which can lead to mechanical stress and/or conductor clearance problems. However, the main source of environmental attack is corrosion of the steel due to salt in coastal areas. Such effects can clearly be seen in any visual inspection.

It is therefore possible to apply HIs to stay wires and to establish a corrosion index that can be related to distance from the coast and the prevailing wind direction.

16.5.1 Cross-arms

Cross-arms can be attacked by the weather, leading to corrosion and rust. However, this is not a significant problem and can be readily picked up by foot or helicopter patrols. In countries that use wooden cross-arms, a visual inspection regime is necessary to determine condition.

16.5.2 Fittings

HV fuses can also be affected by salt pollution, leading to bearings that stick or fittings that do not release under fault conditions. Salt can also cause tracking along a fuse element leading to bulging and failure.

Continual movement due to vibration or general wind effects can affect conductor clamps and insulator attachments. Wear at such fittings can normally only be assessed by visual inspection. Corrosion and vibration can affect the conductor/clamp interface, leading to fatigue problems or poor electrical contact (and subsequent overheating).

In-line tension joints are often fitted at the same span position on each phase. This can look neat but can lead to problems due to clashing in strong winds.

16.6 Foundations

16.6.1 General

Foundations can be stressed by wind/snow/ice loads on the overhead line and weakened by poor design, waterlogged or peaty ground and insufficient soil compression. The foundation of any pole relies on the surface area and strength of soil that gives foundation capability. The use of blocks relies on such soil strength and also the correct size of bolts etc. to transfer the foundation strength to the pole.

16.6.2 Soil strength

It is not possible to use the strength of soil more than once. This may seem like a simple assertion, but the closeness of stay wires at times means that the ultimate design strength is not being achieved. Similarly, blocks and stays should be separated so that the same soil zone is not used. Foundations that are installed using extensive excavation can take several years before soil consolidation gives them full design capability. Heavy rains can reduce the strength of certain soils into the top one metre of ground.

16.7 Pole-mounted equipment (PME)

16.7.1 Lightning

16.7.1.1 General

Pole-mounted transformers (PMT) and vacuum and SF_6 switchgear (PMS) are highly susceptible to lightning strikes. Both require lightning protection to protect against direct and indirect strikes. Air break switches (ABS) and oil circuit breakers (OCB) are considerably less susceptible. Two features of lightning-generated surges are the peak voltage and the surge steepness di/dt, the steepest part of rise of current against time.

16.7.1.2 Peak voltages

The surge peak voltage causes flashovers across external bushings and through internal insulation. In particular, flashovers across the HV bushing can raise tank voltages (on 11 kV lines) to 50 kV or more. This can cause further flashovers across the LV bushings (normal BIL 30 kV) and so send a surge into the LV winding. High peak voltages can therefore cause bushing damage and internal insulation flashovers. Internal flashovers are particularly critical for vacuum and SF_6 switchgear as these units are sealed for life and internal damage can shorten life or even cause failure due to seal damage. Finally, arc erosion of duplex arc gaps (section 16.8) reduces lightning protection capability.

16.7.1.3 Rise time

A high di/dt (10^{12}–10^{13} A/s) means that every conductor must be treated as an impedance and not a resistance. In particular, the surge will drop voltage quite quickly along the HV winding. This means that the voltage between adjacent turns can be far higher than the shellac insulation can withstand and so internal arcing can occur – generally burning out two to three strands (see Figure 18.4). Arcing is also possible between adjacent windings in the case of direct strikes to one phase.

The final effect of a surge passing through a winding is the solenoid effect, where the entire coil is shunted upwards as it tries to eject the core out of the winding (see Figure 18.5).

16.7.2 Snow/ice

Ice can reduce the BIL of bushings, especially when weather conditions have brought particles of pollution into the atmosphere. Generally, however, snow and ice conditions have very little effect on PME.

16.7.3 Wind

Apart from wind-blown debris, wind has little effect on transformers. If sufficient allowance has not been made for temperature variations in jumper connections, then at low temperatures Aeolian vibration can damage these jumper leads.

16.7.4 Pollution

Pollution alone can cause reductions in BIL of the PME bushings. Pollution in snow and ice has already been mentioned, however, the main problem with pollution is salt corrosion of arc gaps and sheath tracking on surge arresters used as lightning protection. These are covered in section 16.8.

16.7.5 Temperature

Ambient temperature has little effect on PME.

16.7.6 Other effects

16.7.6.1 Wildlife

Nesting birds can cause problems for PME in areas where trees are scarce. In addition, falcons and other hunting birds often used PME as vantage or resting points. On taking off, the bird's wings can easily short out the HV leads; this can also occur for other birds, such as owls. Mitigation in areas severely affected is by the use of insulating covers. These will also protect against cats and squirrels.

16.8 Arc gaps and arresters

16.8.1 Lightning

16.8.1.1 Arc gaps

Arc gaps over insulators, bushings and in CC arc protection devices are there to divert lightning current to earth or equalise voltages between phases in CC lines. The main problem with this operation is that the follow-through current erodes the electrodes. A 1000 A earth fault current can erode an 11 kV PMT duplex arc gap by 20 mm per second. One way to avoid this (Figure 16.3) is to use the crossed rod gap (Figure 16.4) where erosion does not affect the arc gap.

Arc electrodes over string insulators and APDs and PADs are generally built of substantial steel cross-section and do not suffer arc gap erosion. Arc gap operation is affected by the polarity of the lightning strike. Arc breakdown will occur at lower voltages for negative strikes and around 10–20 per cent higher for positive strikes. This is due to the higher voltage required for positive streamer formation at the earth electrode.

16.8.1.2 Arresters

Surge arresters used at OHL/cable junctions to protect HV terminals on PME and the neutral bushing on PMTs can suffer lightning damage in three basic ways:

1 Flashovers due to polluted sheaths leaving arc root damage at the terminals.
2 Frequent operation (in a major storm) or lightning currents above rated value can cause deterioration of the ZnO blocks, resulting in less efficient protection and reduced lifetime.

Figure 16.3 Eroded duplex arc gaps

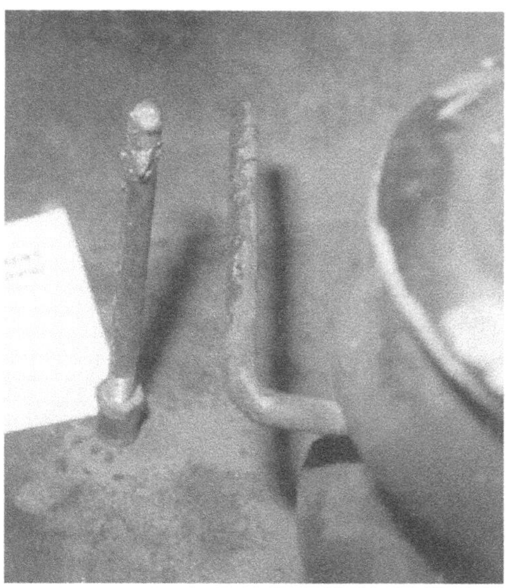

Figure 16.4 Eroded duplex crossed rods still maintain the arc gap

3 Major overcurrent lightning surge can cause immediate failure. Generally, however, the arrester will have held down the lightning surge voltage before failure.

Mitigation is to have an up-to-date surge arrester procurement policy relevant to the local lightning intensity and environment.

16.8.2 Snow/ice

Arc gap operation is not affected by snow or ice except in the case of shorting out by ice formation. Surge arrester sheaths can be affected in the same way as the insulator sheds discussed in section 16.3.

16.8.3 Wind

Wind has little effect on arc gaps or surge arresters, except when on-shore gales bring salt pollution well inland, affecting surge arrester sheaths not chosen to be pollution resistant.

16.8.4 Pollution

Pollution effects can be serious for both arc electrodes and surge arrester sheaths on western facing coasts. Arc electrodes can suffer total elimination due to salt corrosion, especially for duplex gaps. Surge arresters frequently suffer restricted life due to sheaths affected by industrial or salt pollution. The particular variations in coastal environment can determine whether silicon-based or EPDM sheaths or the newer EVA materials can give longer lifetimes. Tracking caused by electric fields over polluted surfaces can reduce impulse and even AC withstand, especially in wet conditions (see Figure 16.1). Shed design, material choice and longer creepage lengths can achieve some degree of mitigation.

16.8.5 Temperature

Ambient temperature has a small effect on arc gap operation, as does relative humidity. Generally, however, these variations are swamped by the natural variation in lightning intensity.

16.8.6 Other effects

16.8.6.1 UV

UV effects on arrester sheaths have caused problems in the past but modern polymeric materials are considerably more resistant to deterioration.

16.8.6.2 Fog

As with insulators, fog has the effect of reducing insulation resistance of surge arrester sheaths. Fine droplets – especially on sheaths already affected by weathering – can reduce surface tracking resistance. Salt fog, in particular, can severely reduce the withstand of silicon-based insulators, even though they recover after the fog has lifted.

16.8.6.3 Rain

Rain and fog can lead to moisture ingress at the terminations – and occasionally through the sheath material – of surge arresters leading to internal deterioration with time.

16.8.6.4 Earthing level and seasonal variation

The earth resistance affects the efficiency of surge arrester protection on PMTs. Depending on the earthing system used, the earth value can vary with the seasons as, although UK rainfall is fairly evenly spread throughout the year, the ability of the ground to absorb rain varies substantially. Earth resistances can be significantly lower in winter and higher, when it matters, in summer, especially if earth mats (usually installed in the top 0.6 m of ground) rather than rods are used. The high earth resistances allow much higher PME tank voltages and hence risk of LV flashover.

16.9 Line design

16.9.1 Lightning

16.9.1.1 General

Lightning is the major cause of faults on overhead lines, accounting for 25–30 per cent of faults at medium voltage and up to 70 per cent at transmission voltages. Line design – incorporating updated protection policies and procurement specifications – represents the best way to tackle this problem. The threat from lightning comes in the following forms:

a number of strikes
b percentage of strikes that cause damage
c consequences of damage.

Although line design can do nothing about item *a*, it can do something about item *b* and particularly item *c*. Of the various components discussed in this section, the most vulnerable to lightning have been PME and covered conductors (CC). The effect of lightning on protection systems (arc gaps and surge arresters) has also been discussed. The line design approach that best deals with the problem is therefore:

1 establish the extent of the problem
2 determine the susceptible areas of the system that need protecting
3 establish cost-effective methods of safely diverting the surge to earth
4 apply protection techniques suitable for the lightning intensity, type of damage and effect of local environment
5 aim to reduce consequences of damage
6 aim to install effective protection in the most effective position.

Following the above principles should provide a protection system that pays for itself in terms of reduced fault levels and improved supply quality. An assessment of current line design lightning protection could grade effectives by looking at:

- standard of protection policies
- standard and condition of protection equipment
- effectiveness of application of protection.

Lightning protection is covered in more detail in chapter 18.

16.9.1.2 Safety

Lines can be designed to withstand environmental effects, but the level of deaths in the public arena due to overhead lines remains extremely high (~15–20 per annum). Some can be mitigated against:

- use of covered or insulated conductors in susceptible areas
- techniques to avoid wildlife problems
- reduction of ROEP due to earth faults
- checks that the network is in a safe condition and within regulatory limits
- line design modifications for particular areas (farmland, building sites, fishing and sailing areas etc.).

With the current emphasis on risk assessment, these could be carried out for typical scenarios as a defence against future deaths or litigation.

16.9.1.3 Snow/ice

Snow/ice loads can cause expensive disruption due to broken conductors and poles and foundation failures. Such disruption occurs due to line components not being designed to match the expected loads or loads being outside expectations. The latter can be addressed on a historical basis with appropriate allowance for climate change. The former has to take account of topographical and geographical location in a way that does not produce an over-complicated policy that is simply not applied in practice.

Condition assessment can take the form of comparing current line standards with those necessary to meet the expected loads in that area. Overall susceptibility can then be assessed.

16.9.1.4 Wind

Forces due to storms, conductor movement and vibration can be assessed. Line design can match wind forces on conductors, poles and PME so that these do not lead to high fault levels. Line design in terms of dealing with vibration relies on knowledge of conductor tension limits throughout the range of ambient temperatures experienced. Long conductor life can therefore be designed in by taking into account factors such as tension, conductor type, terrain and local topography. Galloping is rarely a problem for wood pole lines but needs to be considered for tower lines.

16.9.1.5 Pollution

Pollution generally affects insulated surfaces such as insulators, surge arrester sheaths and CC sheaths. A design choice of insulator material, shed design, creepage length and sheath carbon black content can reduce the effects of pollution.

A second aspect of pollution is the effect of salt corrosion on conductors, stay wires, arcing horns etc. Again, a design choice of material – including possibly accepting a shorter life for some components close to coasts with strong on-shore prevailing winds – may be required.

16.9.1.6 Temperature

Ambient temperature, conductor temperature and joint temperatures can affect line reliability and regulatory clearances as well as lifetimes. Line design can take into account modern conductor technology as well as component reliability so as to optimise network performance. Recent work on conductor ratings and tension/vibration relationships can help in this area.

16.10 Summary

The environment in the UK covers a narrower range than that of most other countries but is also less consistent. Many countries experience certain weather conditions at particular times of the year, but the UK can often experience extremely variable weather in the short term. Although it is not costeffective to meet all that the environment can throw at us, it is certainly possible to focus on particularly susceptible areas costeffectively.

This section has highlighted many areas in the UK where overhead lines are under environmental attack. Other countries facing similar problems have come up with their own solutions. Sometimes it is not necessary to reinvent the wheel or repeat others' mistakes.

An acceptable failure limit has to be decided so that the extent of mitigation or condition assessment techniques can be kept to a costeffective level. Risk assessment should be made for specific scenarios to reduce particular fault-prone areas for safety reasons and to avoid litigation problems.

Chapter 17
Live line working in the UK

17.1 Introduction

This chapter looks at the historical development of working on live lines and illustrates the basics of hot-glove (live line) working on distribution wood pole lines in the UK.

17.2 Historical review

17.2.1 Early development

Live line working in its various forms has been with us for at least 100 years. Initially, long sticks were used to operate switches on energised lines. The ability of long (or hot) sticks to perform maintenance rather than just operate switches on energised lines took a little longer to be appreciated.

The acclaimed father of live line working using hot sticks was T. F. Johnson in the USA. He had developed insulating sticks with fittings by 1910 and by 1914 a full range of tools appeared, including wooden props to lift all three phase conductors and enable cross-arm or pole changes.

The range of operating tools expanded so much that by 1916 spring-loaded tools for jumper work and hot line 'come-a-longs' were in use. By 1918, the Tip Tool Company (eventually taken over by A B Chance now Hubbell) had developed universal tools and sticks with interchangeable heads.

Different techniques developed, where some companies favoured specific devices for holding the conductor and followed the universal tool concept. Others used ropes to hold conductors and interchangeable heads for the sticks. Hot-glove working developed a little later than hot stick, although some basic insulating gloves were around in the early twentieth century. One problem was physical access by the linesman and so hot-glove working did not really take off until the first insulated working platforms appeared in the 1950s.

17.2.2 Voltage levels

Initially, work was restricted to 22 kV or below, but the level was gradually increased as linesmen and companies became more familiar with the technique. Operation was allowed up to 34 kV and then in 1930 up to 66 kV. By 1940, some work had even been performed on live 220 kV networks.

The Hoover Dam suspension insulators (287 kV) were changed live by 1948, and then in the mid-1950s transmission line work up to 345 kV was allowed. The only restrictions that were appearing were that lines had not been designed to be worked live, and so construction features became a limiting factor.

17.2.3 Materials

The first sticks were essentially little different from wooden brooms. Wood, however, especially when wet, is not a suitable material to reduce the current drain to earth through the linesman, and the early operators experienced tingling sensations when connected. Weight was also a consideration, as tiredness can affect work safety levels. Sitka Spruce was a popular choice for many years due to its lightness and good insulation properties, but the basic problem was still the presence of moisture in the wood cell structure.

Moisture problems were overcome by immersing the sticks in cottonseed oil at above 100 °C for several hours. This drove out the water from the cell structure and improved the overall insulation level. Plastic-coated wood was also tried, and then hollow plastic tubes with no wood at all. Hollow plastic and fibreglass tubes came into common use in the late 1940s until the early 1960s. Problems with water ingress from poor fittings eventually led to the development of foam-filled rods where moisture penetration was minimal. This was effectively the lightweight fibreglass hot stick still in use today.

Attention then turned to reducing weight, and aluminium alloy fittings were developed. Linesmen fatigue was also reduced by the development of new procedures and more efficient tools. Procedures are now extremely important in the safety aspects of hot-stick and hot-glove working. The early gloves, as mentioned, were extremely basic and – as with sticks – no electrical testing was carried out. Over the years, new developments were brought in that incorporated testing procedures. The double coloured layered gloves in use today are described later in this chapter.

17.2.4 Standards

In the 1970s, standards and tests were developed to regulate industry work practices. Today, irregular testing is part of any hot-stick operational procedure. One of the main legal requirements in the UK is set by Regulation 14 of the Electricity at Work Regulations 1989.

17.3 Working live

17.3.1 Reasons for working live

When working on live systems there has to be a balance between the advantages gained from working live and the risks associated with it, safety always being the main concern. There are various legal requirements that also come into play when working live. One of the main ones is Regulation 14 of the Electricity at Work Regulations 1989.

The requirements of Regulation 14 (Work on or Near Live Conductors) are to demonstrate that:

a) *It is unreasonable in all circumstances for the conductor to be dead*; **and**
b) *It is reasonable in all circumstances for the person to be at work on or near the conductor while live*; **and**
c) *Suitable precautions (including where necessary the provision of suitable protective equipment) are taken to prevent injury.*

17.3.2 Definition of reasonability

Reasonable is defined in terms of two components.

17.3.2.1 Employee safety incident expectation

This is a statement of the overall level of safety threat to employees undertaking rubber glove working. In a quantified analysis this would be expressed as a number representing the overall expectation of an incident resulting in death or serious injury during the operation being considered. This is not a true measure of risk, although risk calculations would form a significant input.

Employee safety expectation has a large number of components. Examples include:

- number of operations (per year)
- number and types of hazards associated with the operation (including consideration of frequency and consequence)
- risk of death or serious injury (per operation)
- number of employees undertaking the operation.

17.3.2.2 Customer outage expectation

This is a measure of the total disruption to customers associated with the operation being considered, and will also be a combination of several different components. Examples include:

- number of customers affected
- duration
- business sensitive supplies (e.g. high-value processes)
- emotive supplies (e.g. hospitals)
- number of outages (per year).

When talking about customer outage expectations a considerable benefit obtained by working live has to be demonstrated.

The components mentioned above all need to be considered when deciding if it is justifiable to use live line techniques.

17.4 Maintenance using live line techniques

17.4.1 General

There are three main forms of live working, live line tool work (hot sticks), rubber gloving, which can be used on distribution networks, and bare hand (not generally used in the UK).

17.4.2 Live line tool working

Live line tool working is a technique used by linesmen to carry out maintenance on an overhead line system using insulated tools so that the linesman himself stays at a distance to the live conductors.

Most of the modern tools today are manufactured by winding resin impregnated fibreglass fibres onto a preformed core of unicellular polyurethane foam.

In this system the main protection for the worker relies on the insulation of the tool, i.e. its dielectric strength and its electrical insulation properties. For this reason there is a strict regime of inspections and tests to be carried out on the tools at regular intervals.

17.4.3 Rubber glove working

In the early 1900s the first rubber gloves were developed. These gloves were very basic and not subject to any sort of electrical testing. Over the years new developments were brought in until the present day where rubber glove working is an accepted and integral part of working on distribution electricity systems.

When this system of working first appeared the worker would balance on the cross-arm inbetween the conductors and it was not until the 1950s that the first insulated working platforms appeared.

The system in use in United Utilities (UK) today is a combination of rubber glove and live line tool working.

This way of working uses the principle of insulate and isolate: the worker is insulated from the live parts by the use of rubber gloves and isolated from the earth by use of an insulated aerial device (IAD). This does not give total 100 per cent protection and the worker will still need to closely follow all other rules to ensure his safety.

17.5 Stages for a live line job

17.5.1 Stage 1 – preparation onsite

Stage 1 includes all operations prior to the live line activity and includes driving the vehicle to the site and positioning of the vehicle for live working.

Figure 17.1 *Visual inspection of an insulated rubber blanket (courtesy David Horsman, United Utilities)*

The activities following arrival on the working site are:

1 The network control engineer is notified of the rubber glove team's arrival at the site. The auto re-close is to be disabled prior to the work commencing. This is explained in detail at the end of this section.
2 The personal work equipment, other work equipment and PPE is unloaded and a thorough visual inspection of all equipment is carried out. A visual check of equipment is done at the start of the day, i.e. one visual check is completed daily. Each team member is responsible for his own gloves. All work equipment and PPE is to be routinely tested at a regular test interval date (Figure 17.1).
3 The network is checked; this entails a visual check of the physical state of the local network.
4 The person in charge will prepare a job plan.
5 Prework checks are done on the insulated aerial device (IAD). This can be done either onsite or in the depot.
6 The tailboard conference is convened and the job is discussed and documented.

17.5.2 Stage 2 – platform (IAD) raising to initial work position

The tailboard conference is now complete. The buckets are then loaded with the required equipment, and the team members get into the buckets and raise them to the initial work position. All controls are from the bucket. (Note that there is a set of controls within the vehicle.) This stage includes all activities to locate the equipment and operators in the bucket and raise/position the bucket into the initial work area; it includes:

1 loading of the equipment into the bucket
2 boarding of the bucket by operators

Figure 17.2 IAD (courtesy David Horsman, United Utilities)

3 preraise checks (e.g. work equipment, PPE and bucket operation)
4 initial raise of the bucket
5 main raise of the bucket
6 positioning (of the bucket) for work.

The work is watched by a dedicated observer, who is a live line worker and a member of the team; he watches to ensure that the workers are not getting into difficulties and the job plan is being followed (i.e. the actual work is not deviating from the job plan). All the team members are trained to be a dedicated observer. If the dedicated observer has to leave the site, the live working stops (i.e. the job is made safe and the bucket is lowered) until he returns.

Work is completed so that at all times there are at least two levels of protection, i.e. barriers. The work procedures and training develop a method of approach to the first conductor. The primary risk from the raising operation is an electrical hazard from the contact with a live conductor and creating a path for the flow of electrical current. The barriers available to the groundsman and rubber glove team members in the bucket are (Figure 17.2):

- an insulated upper boom
- an insulated lower boom
- an insulated bucket
- an insulated liner inserted in the bucket
- limit of approach; Figure 17.3 demonstrates the principle of limits of approach, the distances are dependent on which part of the body of the worker or vehicle is being measured; the basic principle being always to work at arm's length
- insulated work equipment and PPE (rubber gloves and sleeves, protector gloves)
- cover up to provide a protected work zone.

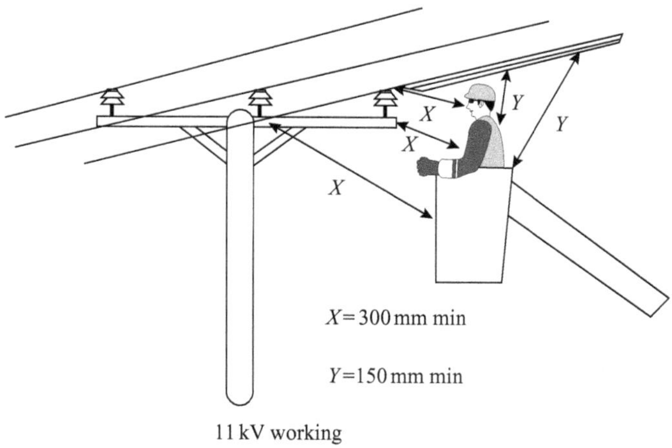

Figure 17.3 The principle of limits of closest approach (courtesy David Horsman, United Utilities)

17.5.2.1 Vehicle movement

During raising (and working) operations, inadvertent movement of the vehicle (rather than the bucket) could result in the uninsulated portion of the boom (below the upper insulated section) making contact with a conductor. In this event, the operator in the bucket would be protected. The barriers available to the dedicated observer are:

- the insulated lower boom
- the path to earth of the vehicle
- keeping a distance from the vehicle.

17.5.3 Stage 3 – work with live overhead lines

Stage 3 includes all work completed from the start to finish of hands-on working, generally, but not exclusively, from the first application of the cover up to the removal of the last piece of cover up. This stage includes:

1 covering up the work area
2 completing work task
3 repositioning of the bucket/IAD
4 picking up of additional equipment; this involves lowering the bucket to ground level
5 removal of cover up (Figure 17.5).

The amount of cover up applied (Figure 17.4) is judgmental and depends upon the work. This barrier has the effect of insulating conductors and any other second point of contact within reach of the operator.

If other live line equipment is needed (i.e. more cover up) the bucket is lowered.

Figure 17.4 *Rubber glove team working on a transformer installation (courtesy David Horsman, United Utilities)*

Figure 17.5 *Rubber glove team using insulated gripall tools (courtesy David Horsman, United Utilities)*

17.5.4 Stage 4 – platform lowering operations

Stage 4 includes all operations to lower the IAD correctly following completion of stage 3. This stage includes:

1 move away from overhead work area
2 main lower

3 final lower (plus docking)
4 post landing checks
5 disembark
6 remove equipment.

17.5.5 Stage 5 – site closedown

Stage 5 includes all operations performed following completion of stage 4; it involves closing the job and leaving the site:

1 pack equipment; this includes preparation of the vehicle for departure
2 check network is left safe
3 check site
4 drive vehicle from site.

17.6 Disablement of auto re-close

There are generally three types of auto re-close features. Auto re-close can be disabled:

- at the site by a lever which is present on the top of the pole
- at a primary sub-station; the auto re-close is disabled by tele-control from the control room engineer.

The third feature is a multi-shot, which requires shorting out to disable.

It is policy to disable auto re-close prior to carrying out live line operations wherever practicable.

17.7 Insulated access (or aerial) device (IAD)

The use of an insulated hoist is necessary when working live to enable the worker to access electrically energised parts of the system.

As this IAD (Figure 17.6) comes into contact with live energised parts of the system, it needs to be insulated itself. Everything, including the booms, hydraulic hoses, basket and jib, is insulated. Because the vehicle is insulated it is subject to a periodic inspection and test; this is done every 6 to 12 months, dependent on the test being carried out.

There are also numerous inspection regimes that need to be carried out by the user on a daily, weekly and monthly basis.

The access device itself is usually mounted on an off-road vehicle of some sort; in this instance it has been mounted onto the back of a Mercedes Unimog. The off-road capabilities of the vehicle are just as important as the capabilities of the booms, as most of the 11 kV network on which these vehicles are used are in the countryside in farmers' fields.

The vehicle also has a lift capability to allow for installation of plant on poles. The amount of weight it can lift is dependent on the angle of the booms.

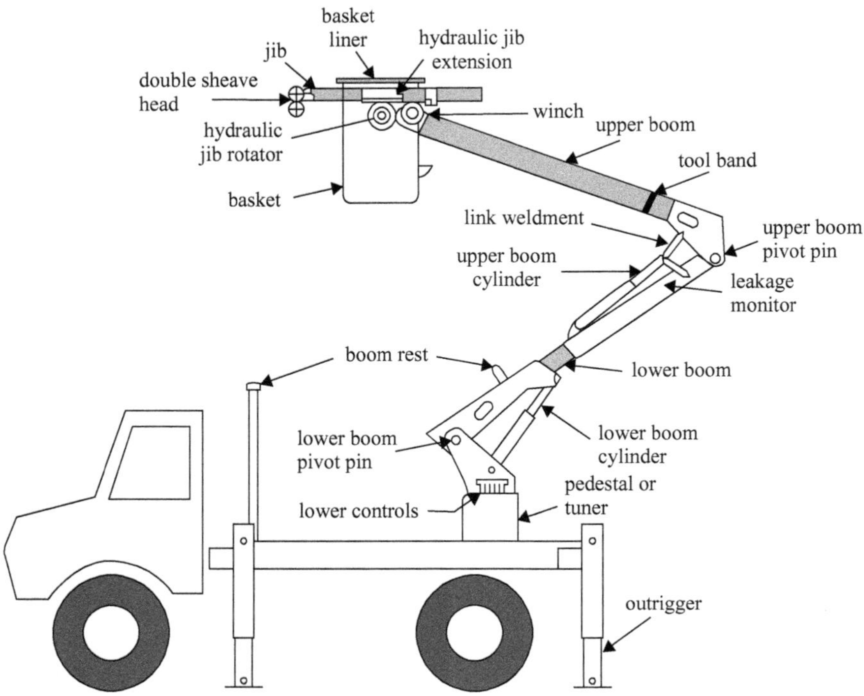

Figure 17.6 Schematic IAD (courtesy David Horsman, United Utilities)

17.8 Personal protective equipment (PPE)

The main PPE that is worn by the worker (in addition to helmet boots and overalls) is a set of insulated gloves and sleeves (these sleeves reach all the way up to the worker's shoulders).

The gloves consist of two parts. The first part is the electrical protection, which comprises natural polyisoprene rubber that is manufactured by dipping the glove a set number of times; in this way different colour rubber can be used for different layers. Generally, each glove has two layers, one that is 20 per cent of the glove and the other that is the remaining 80 per cent, the second one being on the inside of the glove. If the worker, when inspecting the glove, finds a small tear, then by seeing if the second colour rubber is visible or not can tell how deep the tear is and whether or not the glove should be replaced.

The sleeves are made in the same way for the same reason. Both of these items are inspected on a daily basis and electrically tested every three months.

Figure 17.7 shows a glove being visually inspected by inflating it to make it easier to spot any damage to the rubber itself.

Figure 17.8 shows the gloves being electrically tested by dipping them in water, passing an electrical current through the water and measuring how much current leaks through the rubber to the inside.

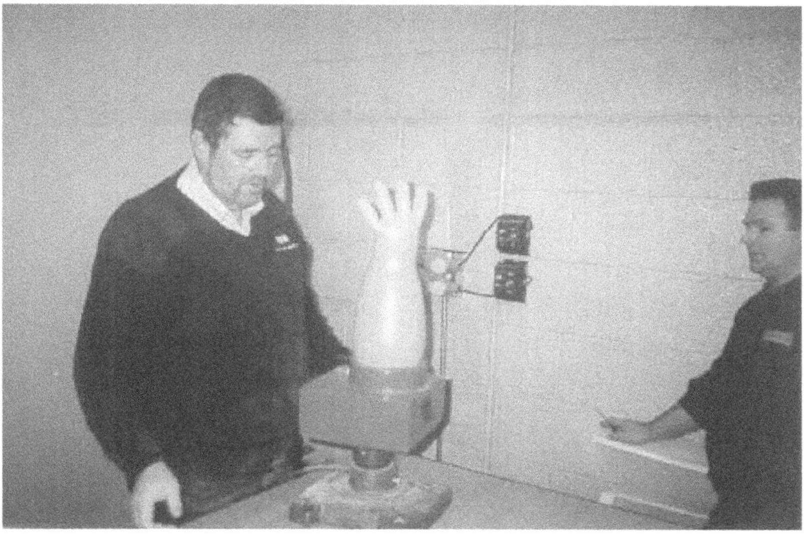

Figure 17.7 *Testing of live line working gloves (courtesy David Horsman, United Utilities)*

Figure 17.8 *Electrical dip test for gloves (courtesy David Horsman, United Utilities)*

17.9 Tools and equipment

There are various tools and pieces of equipment used when carrying out rubber glove working, most of these tools are to enable the worker to handle live conductors without the need to get close to them.

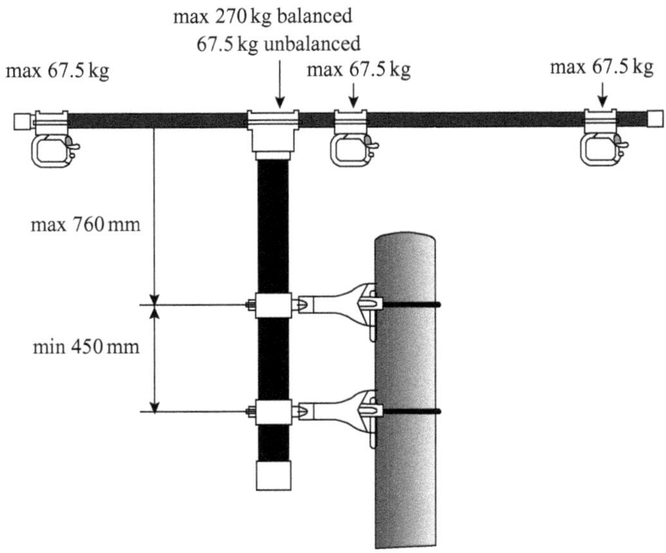

Figure 17.9 Auxiliary conductor boom (courtesy David Horsman, United Utilities)

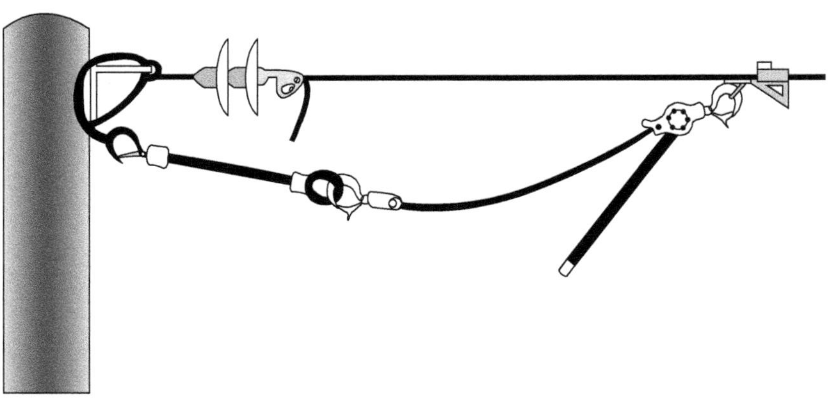

Figure 17.10 Using an insulated ratchet strap (courtesy David Horsman, United Utilities)

Figure 17.9 is a drawing of an auxiliary conductor boom that is used to move the live conductors from the steel cross-arm to enable that cross-arm to be changed.

Figure 17.10 shows an insulated ratchet strap being used to take up tension on the conductor to allow the insulator to be changed.

All of these tools come under the same inspection and testing regime as the other equipment.

Chapter 18

Lightning and lightning protection

18.1 Introduction

In dealing with system outages the utility can often take some pro-active steps by dealing with the original source of the fault, e.g. improving line construction, tree cutting etc. With lightning, the utility has no choice – there is no control over the lightning storm. The only solution is to protect the network to the economic limit of available techniques.

However, in order to protect the network, there is a need first of all to understand the lightning process and how damage occurs. Otherwise, a protection policy may end up as a set of vague or meaningless instructions that can be open to misinterpretation or malpractice in application, as the reasons for them are not fully understood.

This chapter briefly explains what lightning is, how to decide risk levels and how to protect against it. Details of component damage have been provided in chapter 16. This chapter is thus not intended to be a comprehensive guide and more detailed information can be obtained from Cooray [1].

18.2 Lightning effects on overhead lines

18.2.1 General

It is perhaps unfortunate that lightning is not the same all over the world. Although it is extremely common – at any one time there will be around 2000 thunderstorms [2] active somewhere in the world – it is essentially a random process in the short term. In equatorial regions such as Singapore, there are 180–200 days each year when lightning storms occur. This is based on the original way of measuring lightning activity – whether it was observed or heard by a human observer. In the UK the lightning intensity is only one tenth of this level. Predictions of lightning activity are possible over the longer term on a theoretical as well as historical data basis. It is this long-term lightning pattern that is relevant for predicting lifetimes of overhead lines and equipment. The short-term random nature of lightning, however, cannot be neglected and the relevance of this in terms of lightning protection will be discussed later.

Generally, in the southern UK, lightning strikes occur at rates up to 2 str/km^2/yr (10–20 thunderstorm days/yr) with mean lightning currents of 10 to 20 kA. These strikes are mainly what is known as negative lightning with multiple strokes and short rise times (the current rises from zero to the peak value in less than 1 μs) and occur in the summer months. In the northern and western parts of the UK there is a greater percentage of positive lightning strikes. These have different characteristics, often being single strikes with slower rise times but with much higher current levels – up to 500 kA having been recorded for a single strike. In these areas severe lightning activity can occur during the winter months.

18.2.2 The thunderstorm

The thunderstorm can develop as a single spontaneous cloud or as a line of storm clouds formed by the rising of warm moist air along a weather front. It is formed by rapidly rising, warm air masses caused by locally unstable atmospheric conditions. Movement of water and ice particles within the cloud generates static electric charges at the top and base of the cloud.

The charge on the base of the cloud induces an opposite charge on the ground and the resultant discharge is the lightning strike. The sudden heating (and collapse) of the air column caused by the passage of lightning current generates a compression sound wave (thunder) that travels at around 300 ms^{-1}. As the lightning strike occurs within microseconds, the time difference between visual observation of the strike and hearing the thunder can accurately determine the distance of the discharge from the observer.

18.2.3 The negative lightning strike

The negative strike originates as a barely visible, slowly moving leader from a cloud. This leader move in jumps of 50 to 100 m at an average speed of 0.15 m/μs. These jumps are probably preceded by the creation of short lengths of ionised path. This leader will make contact with one or more streamers that rise up from objects on the ground. The leader will have a general tendency, as with any arc, to be attracted first of all to any sharp points (e.g. towers, trees, poles) and then to any objects with small radius of curvature (e.g. horizontal rods, overhead line conductors) and last of all to a flat surface (the ground). The negative strike to a line will cause an overvoltage surge to travel in both directions independent of the actual direction of power flow in the line. The most common form of protection is by arc gaps at cross-arms or pole-mounted equipment. The overvoltage surge will break down across the arc gap (the actual process will be described later) causing a discharge either between phases or phase to earth (or transformer/switchgear tank).

18.2.4 The positive strike

The initial formation of a positive strike is the same as that described for a negative strike in the previous section except that it is now thought that the discharge comes from the leading upper part of the cloud rather than the lower part (for the negative strike). The discharge travels at a slower speed in the lower density upper atmosphere and so the current rise time is longer for positive strikes. However, it has been well known for nearly 40 years that arc gaps on overhead lines react differently to positive and negative strikes [3]. The present protection system is based on the more common negative strike. A positive strike can break down at lower voltages than a negative strike and thus the gap also has a lower withstand voltage for positive surges. This implies that breakdown between phases can occur more easily with a positive strike.

Most data on the effects of positive lightning are from northern Japan [4], where the ratio of positive to negative strikes was measured as 1:2 as opposed to 1:9 generally (and in central and southern UK). In the northern and western UK the number of positive strikes can exceed the number of negative strikes (information from EA Technology Ltd lightning detection system archived data [5]).

A negative lightning strike nearby or directly to the line will generate an overvoltage surge on the line. This will have a rapid rise to a peak value followed by a slow fall-off or tail.

Data so far indicate that positive lightning has a much slower rise time to peak voltage – over ten times slower than that for negative strokes – and a much slower decay to 50 per cent of its peak value – nearly 20 times longer than for a negative strike. This results in the positive lightning strike having a much greater total coulomb charge than a negative strike. Japanese figures indicate levels of charge ten to 20 times higher for positive lightning, and this is borne out by EA Technology data [5].

Thus, positive strikes may need to be treated differently when considering the susceptibility of overhead lines to damage and methods of protection.

18.3 Theory of the lightning strike

18.3.1 Introduction

Most of the early theoretical work on the coupling of a lightning strike to an overhead line was done for transmission lines. However, the basic theory still applies to the lower heights of wood pole lines and is indeed more relevant as the lower basic insulation level (BIL) of medium voltage lines means that they are more susceptible to lightning damage.

Most lightning generated overvoltages on overhead lines are induced from a strike to the ground or other object near the line. The theory of induced overvoltages was laid down by Rusck [6] and the distribution of current in the strike given by Anderson and Eriksson [7]. Induced lightning strikes can cause some damage, but direct strikes to one of the phase conductors are generally of more concern. The theory of the attraction of the lightning strike to the actual line rather than the ground or other nearby object is based on two alternative models: the electrogeometric model [8–10] and Eriksson's model [11].

18.3.2 Direct strikes

In the UK, most lightning current levels are in the 10 to 30 kA range, although in some areas in the north and northwest average levels can reach 80 kA. As this lightning current surge hits the line the overvoltage level in the line will be determined by the current level times the total fault impedance at that frequency. It can be seen, therefore, that multiple insulation flashovers (between phases and to ground) will occur until this multiple of current and impedance reduces to below the BIL of the line. In the case of the normal level of lightning current and the surge impedance of a wood pole MV line, being around 300 to 500 Ω, virtually all direct strikes will generate overvoltages well above, typically, the 95 kV BIL of an 11 kV line.

In the electrogeometric model [9, 10], lightning that would have struck the ground a distance l from the line will instead be attracted to and strike the line directly if $l < l_d$ where:

$$l_d = \sqrt{(r^2 - (r - h)^2)} \text{ metres} \tag{18.1}$$

and

$$r = B \cdot I^{0.65} \tag{18.2}$$

where l_d is the horizontal attraction distance (m), r is the distance from where the strike diverted (m), h is the line height (m), I is the return stroke current (kA) and B is a factor that varies with line geometry.

For a horizontal three-phase line, B is normally given the value of 10.

Eriksson [11] gives a different expression for l_d:

$$l_d = 0.84 I^{0.74} h^{0.6} \tag{18.3}$$

The frequency of direct strikes can be evaluated from these expressions if the strike density in the area is known. If a line has a length, L, with phase spacing, d, then lightning would strike the line rather than the ground if it is within an area, A, around the line given by:

$$A(I) = 2L[l_d(I) + d] \tag{18.4}$$

Historical data should give the strike density N_g str/km^2/yr and the probability of any particular current, I, i.e. $f(I)dI$. The average number of strikes to the line per year will be:

$$N_d = 2 \cdot L \cdot N_g \int (l_d + d) f(I) dI \tag{18.5}$$

If the lightning current has a mean value of 34 kA, then a line with 10 m high poles and phase spacing of 1 m will suffer between 8.7 and 10 direct strikes per 100 km circuit length per annum if the strike density is 1 str/km^2/yr. This is based on the simpler expressions (18.1) to (18.4). Expression (18.5) will give a more accurate picture.

18.4 Indirect strikes

The Rusck theory [6] is basically simple but gives results in good agreement with experimental data. The maximum induced voltage will occur at the nearest point of the line to the ground strike. Once the induced voltage for a particular strike current value and horizontal distance from the line has been evaluated, then this can be compared to the BIL of the line or equipment. For any particular medium voltage line (or LV spur) the BIL value will therefore determine a corridor within which a strike to the ground will generate damaging overvoltages on the network. The size of this corridor will be related to the lightning current value and to whether the strike is positive or negative.

At this stage, the simplest approach is to take the long-term historical mean of the lightning current and calculate the corridor size. If this is taken with the mean strike density as described in section 18.3.2, then the number of damaging indirect strikes can be calculated for the lifetime of the line and the level of protection determined.

For lines demanding high security then a more rigorous analysis should be performed. This should include the historical spread of lightning current values integrated over the strike rate to give the line's overall susceptibility. The likelihood of particularly damaging levels of activity can then be matched with a level of lightning protection based on the security needs of the line.

18.5 Standards

Standards are an essential part of engineering and scientific life and they are especially important in lightning protection. Standards initially define the various types of lightning surge that occur on overhead lines and cables in terms of the time to peak voltage and decay to 50 per cent peak voltage, so that all HV laboratories world-wide can perform identical standard tests on equipment and protection systems.

Standards also enable consistent methods to be applied and frequent revision can allow for new techniques and knowledge to be used to keep these methods updated. Standards are normally set and updated by experienced engineers with in-depth knowledge of lightning, protection systems and equipment and/or architectural design. It is therefore well worth devoting a section to their study.

Standards provide authoritative guidance on the principles and practices of lightning protection for a wide range of structures and systems. They also help to define what is a quantitative risk and whether protection can be justified. Although generally not mandatory, compliance with current standards can be a substantial defence in contractual problems. This chapter generally follows the standards set in BS 6651 (1999) and IEC 61662.

18.6 Risk management

Risk management follows a specific pattern:

1 identification of the lightning threat
2 the frequency of the risk event
3 the consequences of the risk event
4 establishment of a tolerable risk level

5 if required, specification of protection measures to reduce the risk to a tolerable level.

If the risk from direct lightning is R_d and indirect lightning is R_i, then the risk of a total of events, x_i is:

$$\sum R_x = \sum R_d + \sum R_i \tag{18.6}$$

and

$$R_x = N_x P_x \partial_x \tag{18.7}$$

where N_x is the event frequency, P_x is the probability of damage or injury and ∂_x is the consequential effect of the damage or injury.

So the overall probability of a damaging event will be:

$$P = k_x P_x$$

where k_x is the reduction factor obtained by applying lightning protection. So, if R_t is the tolerable level, the end result required is:

$$R \leq R_t \tag{18.8}$$

18.7 Lightning protection systems

Lightning protection systems (LPS) follow basic principles to:

1 protect against lightning electro-magnetic pulse (LEMP) or surge from direct and indirect strikes
2 protect against transient pulses (TP) damaging electrical and telecommunication systems or using these to enter other areas and so damage people or sensitive equipment
3 design an earth electrode system to dissipate the current into the ground safely
4 recommend measures to limit magnetic fields and earthing (earth mats)
6 recommend surge protection devices
7 evaluate insulation coordination.

This chapter will expand on these areas and explain the concept of lightning protection design.

18.8 Calculating risk

18.8.1 General

If the calculated risks are sufficiently high, then lightning protection may be required – if it is both effective and economic.

The approach detailed here is based on IEC 61662. This considers evaluation of the risk to a structure, people, installations or equipment within a structure or in some way

Table 18.1 Probability of death from various causes

Activity	Chance	Annual probability
Smoking ten cigarettes/day	1 in 400	2.5×10^{-3}
All accidents	1 in 2000	5×10^{-4}
Traffic accident	1 in 8000	1.3×10^{-4}
Leukaemia	1 in 20 000	5×10^{-5}
Drowning	1 in 30 000	3.3×10^{-5}
Poisoning	1 in 100 000	1×10^{-5}
BS 6651 tolerable level	1 in 100 000	1×10^{-5}
Natural disaster	1 in 500 000	2×10^{-6}
Struck by lightning	1 in 2 000 000	5×10^{-7}

connected to it. This includes mechanical and electrical damage, failure of equipment, voltage differences causing step and touch potential injuries and fire damage.

18.8.2 Risk

Risk, R, is defined as the annual loss due to lightning – essentially the probability of loss over one year in time. But it is necessary to be realistic about the probability of loss. There are many areas of life where calculated risks are taken or just ignored because the risk is not considered important or relevant or because nothing can apparently be done about it. Table 18.1 gives some typical examples of the probability of death. These are taken from the Australian standard AS 1768.

All the three main standards (AS 1768, BS 6651, IEC 61662) cover four areas of risk:

1 loss of human life
2 loss of a public service (e.g. electricity supply)
3 loss of cultural heritage
4 loss of economic value.

A tolerable risk value for human life is set at 10^{-5}, and service and heritage are put at 10^{-3}. Loss of economic value is not specified as it depends very much on the particular situation. Modern standards place a greater emphasis on saving life than on preserving cultural heritage – these used to be classified as of equal importance.

18.9 Damage

In this section, just three types of damage will be considered:

1 injury to people (step and touch potential, side flash)
2 major mechanical damage (e.g. fire, explosion)
3 failure of electrical equipment.

18.9.1 Risk components

The total risk due to lightning will be made up of several components, all of which must be summed to make up the lightning risk per annum.

18.9.1.1 Direct strikes (R_d)

Direct strikes can cause step and touch voltage gradients (risk R_h), thermal effect and mechanical damage (R_s) and overvoltages affecting electrical equipment (R_w). Overvoltages can be transmitted through conducting paths to create high touch voltages in other areas (R_g). Surge travel along these paths can also create arcs, either deliberately or at bad joints (R_c). Surges along overhead lines will also affect electrical and electronic equipment (R_e).

18.9.1.2 Indirect strikes (R_i)

Indirect strikes can also generate overvoltages by the magnetic field associated with the strike. These can affect electrical equipment directly (R_m) and also indirectly by transmission of an induced overhead line surge into electrical equipment (R_e).

Consider each component risk, R_x, which has a probability of damage of P_x and a frequency of N_x. The damage may have a high or low consequential effect and this is represented by δ_x. In this case, the risk component is evaluated as:

$$R_x = N_x \cdot P_x \cdot \delta_x$$

In terms of lightning, the factors N_x and P_x can be readily evaluated. Frequency of events can come from historical data and the probability of damage will come from the number of those events likely to exceed a certain damage threshold. δ_x is evaluated in the following three generic examples.

Loss of human life

If n is the number of likely victims of the strike and n_t the total number of people normally involved in the area, then δ_x, the damage factor, is obtained from:

$$\delta_x = \frac{n}{n_t} \cdot \frac{t}{8760} \tag{18.9}$$

where t is the time (in hours) that people spend in the area per year (there are 8760 hours in a year).

Loss of public service

Along the same lines, if n is now the number of consumers off service and n_t is the total number of consumers served, then δ_x is given by the same expression (equation (18.9)), where t now is the annual loss in service (in hours).

Economic loss

This is slightly different. δ_x now is simply the ratio of the annual loss of equipment and other associated losses, c, compared to the total value of equipment and services provided, c_t. So:

$$\delta_x = \frac{c}{c_t} \tag{18.10}$$

So, having identified all the types of loss, the frequency and probability and the damage factor, it is now possible to identify each risk component R_x. All these must be added together to give the overall risk, R. The risk that can be tolerated, R_a, must be established. Lightning standards generally state a tolerable level of maybe 1 in 10^5. This may be higher or lower depending on whether or not special factors are involved.

If $R < R_a$, then protection is generally not required. If $R > R_a$, then protection should be applied. One last point – this whole procedure should be repeated once protection has been decided upon. The new component risk levels with protection included should again be calculated and added. Hopefully, R will now be less than R_a but, if not, then a further degree of protection is required.

18.10 Protection of structures

18.10.1 General

In the electricity industry the structures that are of most concern are sub-stations (which may be open or enclosed in buildings) or overhead lines and cables. These will be covered later but, for now, a brief guide to lightning damage to buildings.

The most vulnerable areas are those on the upper parts of the building:

- aerials and other roof structures
- pointed apex roofs
- spires
- gable ends (roof ridge section)
- outer roof corners (especially for flat roofs).

Not quite so vulnerable are:

- edges of flat roofs
- slanting edges of gable ends.

And least vulnerable are:

- vertical edge of the building.

The reason for vulnerability is height, plus the electric field intensification associated with exposed points and corners. With this in mind, a lightning protection terminal should be designed to cover these vulnerable areas.

Further vulnerable points can be created by poor design of the down conductor to earth from the lightning terminal. This can result in side flashes through the building fabric to electrical circuits and equipment within the building.

Later, the rolling sphere method of designing lightning terminals will be covered. As it can be difficult or unaesthetic to have a single high terminal, the use of multiple

Table 18.2 Radii of rolling spheres

Protection level	Sphere radius (m)		Minimum strike current (kA)
	a standard radius	a_i increased radius	
I	20	60	2.9
II	30	60	5.4
III	45	90	10.1
IV	60	120	15.7

terminals is recommended. These must be spaced closely enough to reduce the risk of lightning strike on vulnerable areas between the terminals.

Low earth resistances are essential. In most cases, values of <10 Ω, but in high priority areas, <1 Ω may be required. In exceptional circumstances, where risks are high, earth resistances below 0.1 Ω may be required. To reduce electric surge impedance in the down conductors, bends should be avoided due to their inherent high inductance at lightning frequencies but, if necessary, they should be slow and gentle. Right angle bends should never be used due to the high inductance created.

18.10.2 Rolling sphere method (RSM)

The rolling sphere method is basically a simple technique to establish whether a lightning terminal is in the right place and of sufficient height to give the zone of protection desired. The lightning strike distance (determined by the strike current) is set as the radius of the sphere. The aim is to ensure that the shortest distance between the lightning downward leader tip and the structure is the lightning terminal.

In theory, a sphere of radius a (given in Table 18.2) is rolled over the structure. Any sections of the structure touched by the sphere are considered vulnerable to a strike. In this context, it is required that the sphere makes contact first with the lightning terminal or its interception surface.

The increased radius, a_i, is used when considering a large plane area that does not give any field enhancement. Protection level I is the highest level of protection, as it covers lightning strike currents down to 3 kA. According to empirical data from Cigré, level I has an interception efficiency of 99 per cent, compared with 91 per cent for level III. This is based on the proportion of strike currents below these levels and the fact that the shorter lightning strike distances of the lower currents can get through the zone protection distances designed on higher currents.

The Australian standard uses a different version of the RSM, which is designed to give weight to the enhanced field effects of corners and edges. The standard, and tighter, sphere radius, a, is used for structures rising above a plane surface, whereas the increased radius is used for protection of large plane surfaces where there is no field enhancement. In the Australian standard, a plane surface is defined as one where there is nothing higher than 0.3 m.

Rather than physically using a sphere (or flat circle) over a cross-sectional drawing, it may be easier to calculate the size of lightning terminal required to protect a particular area. If r is the distance from a terminal of height h (above the local flat plane – which may be a building roof) to the equipment or roof edge to be protected:

$$R = (2ah - h^2)^{1/2} \tag{18.11}$$

where a is from Table 18.2. For a large flat plane a_i can be substituted for a to give R. A large flat area can be split into a square grid of unit side length d. It can be seen that if one air terminal is set to cover each grid section, $d \leq r \cdot \sqrt{2}$.

18.11 Protection techniques

18.11.1 General

An overhead line network consists of many different and often interacting components. The fault rate at sub-stations is dependent on the several kilometres of lines and cables leading into the station, as these are the major source of lightning-generated overvoltage surges. The severity of faults on pole-mounted equipment can be dependent on the closeness to a sub-station and levels of follow-through current. So, although this chapter looks at protecting overhead lines, sub-stations and covered conductors independently, it is thus necessary to take an overall view of protection strategy.

Arc gaps respond better to flattened wave forms whereas surge arresters respond rapidly to the sharp-fronted surge, although the voltage held across the arrester will be slightly higher for the sharp-fronted than for the flattened wave. It can be seen therefore that the flattened wave form of induced strikes or positive lightning may allow arc gaps the longer time they need in which to operate and still protect pole-mounted equipment.

18.11.2 Arc gaps

Arc gaps are a traditional method of diverting the lightning-generated overvoltage surge to earth. The discharge across the gap depends on the local presence of electrons or ionised gas. Free electrons are present everywhere and therefore the time for a breakdown can be reasonably accurately predicted. Ionised gas will still be present in the milliseconds after a discharge has occurred in the local area of the discharge. This is why once an arc gap has operated due to an overvoltage surge, it will break down again immediately under normal line voltage which it would otherwise normally withstand. Arc gaps can protect from slow rate-of-rise surges. However, they will transmit fault current after an initial discharge and this can cause insulator or bushing damage, electrode erosion and circuit breaker operation (see chapter 16 for component damage).

18.11.3 Surge arresters

Arresters come in several classifications and it is unfortunately fairly common for them to be ordered without a direct knowledge of their application or environment, and to be installed in unsuitable positions. This leads to poor and of course uneconomic protection efficiency. Utilities should look closely at their need for both procurement guidance and a full protection policy relevant to their own requirements and environment.

18.11.4 Overhead earthwires

The earthwire, when used as a lightning protection system, is supposed to collect all the lightning strikes that would otherwise have struck the phase conductors. It is based on the assumption that the downward leader descends essentially vertically and that it will meet up with an ascending streamer from the earthwire rather than a phase conductor. This is fine for large strike currents with long jump distances. However, leaders with low strike currents can get closer to the phase conductors by striking from the side of the line. This is known as shielding failure. As there will always be a spread of strike currents, there will always be a percentage risk of shielding failure. This can be reduced by having negative angles between the phase and earth conductors as with twin earthwires on steel portal structures. However, the most common UK 132 kV tower designs all have just one earthwire and a positive angle of 45°. This will allow a high percentage of low-current strikes to penetrate the earth shield. At this angle a 10 kA strike will have a 40 per cent chance of striking a phase rather than earthwire. In the southern UK the strike currents are low and hence the risk of shielding failure is high. This has particular relevance for tower lines feeding sub-stations.

18.11.5 Tower footing impedance

The weakness of a single overrunning earthwire as a lightning protection device has been shown. However, even if this system was improved and became 100 per cent safe, the tower line would still suffer from lightning strikes if the tower earth impedance were poor. The problem here is back flashovers. When the earthwire or tower is struck it will rise in potential to a value depending on:

- the steel tower impedance
- the footing impedance
- the strike current.

It is important to remember that it is the impedance at megahertz frequencies that is being considered – not the DC or 50 Hz resistance. Skin depth considerations mean that steel girders have much higher impedances at lightning surge frequencies than at power frequencies. The tower can easily rise to 2 or 3 MV. If the voltage rise of the tower exceeds the insulation level of the suspension insulators then there will be a flashover from the tower arm to the phase conductors (back flashover) – often on more than one circuit. The problem here is that the breakdown will put a sharp-ended overvoltage wave onto the phase conductor. This will be a problem to equipment insulation if the line is close to a sub-station.

18.12 Conductor type and line configuration

18.12.1 General

Generally it is not the lightning strike that does the damage – indeed most damage is done to 11 kV networks by lightning that strikes the ground up to 50 or 100 m away from the line. Both direct and indirect strikes generate overvoltage surges that travel in both directions (load and scale) along all the conductors they can find. A lightning strike occurs when a downward tracking leader makes contact with one or more upward tracking streamers from objects near, on or even underneath the ground [2].

18.12.2 Overhead line conductors

Direct lightning strikes generate high voltage surges that can lead to damaging arcs. Most arc damage will occur on the conductor close to the support insulator or to the insulator itself (the nearest low BIL point), not at the source of the actual strike (unless it is a covered conductor). The low BIL point at which the breakdown occurs may not be at the first pole. Due to the elevation of voltage on the two other phases (induced by the strike) as well as the struck phase, the lowest BIL point may be at an earthed cross-arm. The breakdown will raise the voltage of the earthed cross-arm well above earth potential, especially in the case of poor grounding. There will, therefore, in the absence of lightning protection, be a substantial arc discharge across the insulators. Arc initiation needs a very high voltage but very low voltages – well below normal line voltages – can sustain arcs. A phase–phase arc will travel along the line in the direction of the load and possibly to equipment on the pole but will not extinguish until circuit breaker operation occurs.

In the case of unearthed poles, there is essentially a duplex gap between adjacent phases with the cross-arm as the floating middle electrode. There is also an effective duplex gap formed by one pin or pilot or post insulator and a stay wire insulator at angle poles. In the absence of lightning protection, breakdown is likely to occur at these points. In the case of an indirect strike, the induced voltages on all three phases will be very similar and there is unlikely to be any breakdown at unearthed poles although there may be breakdown at stayed poles. At an earthed pole there may be a breakdown as for a direct strike but because the voltage levels will be lower and the rise times slower, severe damage is less likely to occur.

18.12.3 Bare wire

Streamer generation depends on the ionisation level around the object. A small bare wire conductor will generate streamers readily due to its small radii and therefore higher electric field gradient locally. However, this type of conductor will also reduce the travelling surge more quickly due to corona losses. Bare wire lines will therefore have a higher incidence of lightning strikes but a greater loss rate when struck.

18.12.4 Covered conductor

A covered conductor will act like a larger conductor with a reduced surface voltage at a larger diameter due to its carbon content sheath. These effects will reduce the ability to generate streamers. This implies a narrower collection width and so less direct strikes compared with bare wire. This has a greater implication for 33 kV lines than 11 kV as only ten per cent of 11 kV damaging strikes are direct compared with 50 per cent for 33 kV lines. On the other side of the coin, the higher corona inception voltage of the covered conductor will cause it to have a reduced ability to reduce surge voltages by corona loss. Covered conductors are discussed in more detail in section 18.13.

18.12.5 Overhead cable

The only overhead cable currently in use in the UK is the Swedish Axces and Excel cable. This has three cores individually surrounded by XLPE insulation but with one overall earthed sheath and an external LLD PE insulation. The overall larger diameter, the lack of an external electric field and the high BIL (400 kV) will make this cable virtually immune to induced strikes and considerably less attractive for direct strikes compared with bare wire.

18.12.6 Underground cable

The voltages in underground cables generally fall with increasing cable length due to attenuation. There is a difference in surge impedance between long and short cables as well as different reflection patterns that occur within cables of different cable lengths. The original ACE 55 report introduced the concept of a self-protecting length of underground cable. This concept is now generally discredited in light of improved knowledge of lightning and its effects. The advent of the cruciform surge arrester cable end fitting has also made cable protection easier and more economic.

18.12.7 Protection strategies

There is no real need for a protection strategy for bare wire lines – the protection needs to be focused on any pole-mounted equipment or sub-stations. Covered conductors are different – these do need protection for the conductor itself and the equipment. The overhead cable should have no overhead/underground cable junctions and the low susceptibility to lightning should make protection of this cable unnecessary. However, any connection joint between other types of overhead conductor and an underground cable does need protection.

18.13 Covered conductor protection

18.13.1 General

Conductor clashing, blown or vandal flung debris, fallen trees etc. can cause short- or long-term faults. Contact to the line by sports equipment such as fishing rods and by birds can also prove dangerous to third parties. Conductors covered by an XLPE sheath can reduce or eliminate some of these problems but bring in a problem of their own. Covered conductors (CC) lines are susceptible to lightning strikes because any phase–phase arc generated by a direct strike will not travel along the line as on a bare line, but will stay located at the struck point. Conductor burn-down is therefore possible. The use of CC lines therefore requires a further co-ordinated approach to lightning protection as a whole for the line and its ancillary equipment, in particular PMTs.

18.13.2 Direct strikes

If there is a direct strike to a bare line, then the arc root on the line will move due to magnetic effects. This means that subsequent strokes within the same strike will most likely not strike at the same spot, but may be displaced by several millimetres.

However, if the line is a covered conductor, the arc root will remain in the same spot, constricted by the hole drilled through the sheath.

The conductor at the strike point will suffer a direct energy transfer from the stroke and also joule heating from the current. In the short time scale radial heat loss will be very small and the major heat gain will be from joule heating. Taking the thermal conductivity, specific heat and density of aluminium, a 30 kA strike will generate sufficient heat to melt the metal at the struck point and raise the temperature on the opposite side of a 50 mm^2 conductor to over 200 °C within 0.2 s. If allowance is made for a 2.3 mm XLPE sheath then the temperature at the centre of the conductor will not rise above 120 °C. The initial strike therefore will leave a 2–3 mm diameter hole in the sheath and probably melt part or all of one stand, possibly reducing the strength of others by local annealing. In this, the damage will be similar to that experienced by bare lines – the problem for CC lines being water ingress through the hole.

At the same time as the strike, streamers will be generated from the adjacent phase conductor. These will make small holes (<1 mm diameter) in the sheath and start to ionise the air between the struck and non-struck phases. After around 4 μs (depending on the conductor size and spacing) this streamer will develop into a full-blown arc that will then draw phase–phase fault current. Major damage will then occur, possibly leading to failure of the struck conductor. This will suffer the worst damage, as the aluminium will already have a molten surface.

The aim of protection is to avoid this phase–phase flashover. No protection (other than a shield wire) can avoid the original damage to the conductor.

18.13.3 Indirect strikes

Indirect strikes will not cause sheath damage unless there is an arc generated at earthed cross-arms. The need here is for a form of arc gap or arrester protection that either allows a safe arc generation or avoids the arc totally. The problem then passes onto pole-mounted equipment.

18.13.4 Damage to associated equipment

On a CC line that has been struck by lightning, the struck phase will rise in voltage by several 100 kV, and the other two phases will have lower-level induced overvoltage surges. These surges will travel out in both directions until the low BIL point is reached. Normally this will be an unearthed pole. In the absence of protection a flashover is likely to occur over the line insulator between the struck phase and the cross-arm and then from the cross-arm to one of the other phases. If there is no break in the conductor sheath at the first pole then this breakdown may take time to develop and the arc may establish instead at the struck point (mid-span). However, if breakdown at the pole top does occur then the arc will cause insulator damage due to thermal shock, discharge effects or tracking.

If the travelling overvoltage surges from direct or indirect strikes reach an earthed pole containing pole-mounted equipment, then the equipment represents a low BIL point and damage can occur exactly as on a bare line. On a CC line there are actually more arc gaps – at APDs, PADs and stay insulators. The existence of these arc gaps means that the overvoltage surge level will be reduced before it reaches the pole with equipment.

18.13.5 Protection strategy

In its simplest form, the CC line will require arc gap protection between the conductor (using an insulated piercing connector or IPC) and the cross-arm. Two types are used:

1 Arc protection devices (APD), which consist of IPCs with simple electrodes attached and mounted on each phase on both sides of the pole.
2 Power arc devices (PAD), which consist of an IPC with an attached electrode and a second electrode, mounted on the cross-arm. These are required on each phase but only on one side of the pole.

PADs operate under any phase spacing and any level of fault current. APDs only operate on phase spacings below 700 mm and fault currents above 1 kA. Both devices require circuit breaker operation to eliminate the arc.

Triggered arc (or spark) gaps (TAG or TSG) are a more expensive device, usually mounted on the pole with connections to the conductor and to earth (Figure 18.1).

Figure 18.1 Triggered arc gap

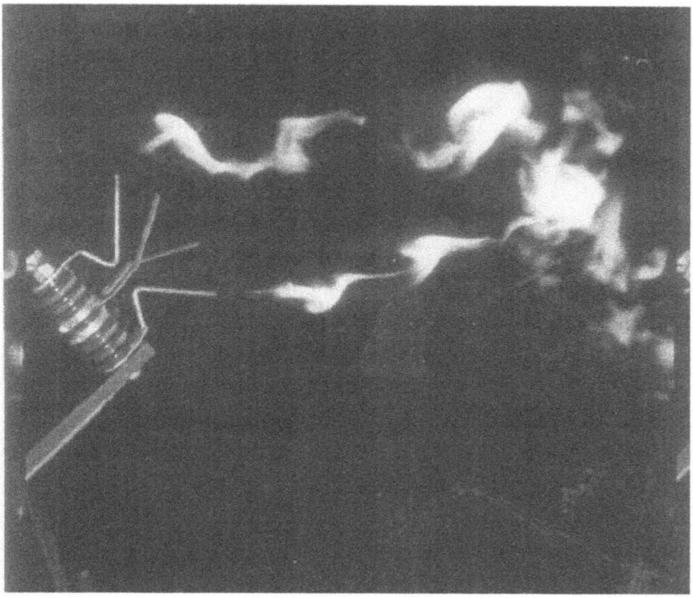

Figure 18.2 TAG showing the moment of extinction of the arc (courtesy EA Technology Ltd)

They are designed to be self-extinguishing (Figure 18.2) and so do not require circuit breaker operation.

Surge arresters can be used mounted between each phase and the cross-arm. These are the ultimate in maintaining supply quality during a lightning storm but are the most capital expensive option.

Table 18.3 Response time of a duplex arc gap (μs)

Voltage (kV)	Negative strike (gap)		Positive strike (gap)	
	25 mm	45 mm	25 mm	45 mm
100	1	2.5	1.6	4
150	0.7	1.4	0.8	1.1
200	0.5	0.9	0.5	0.8
300	0.35	0.5	0.3	0.4

18.13.6 Arc gaps

The normal arc protection device (APD) on a narrow construction intermediate pole relies on an initial arc from the trace wire across the pin insulator and onto the cross-arm. A second discharge is necessary across the pin insulator of the neighbouring phase, before the arc can then establish itself across the APD arcing horns. If the arc discharge does not make this transition to the APDs then the follow-through current may damage the insulators.

The speed of this arc development depends on the voltage of the lightning surge and the level of the follow-through current. From Table 18.3 it can be seen that for voltage levels only just above the BIL of the system, there will be a delay of at least 2.5 μs before breakdown between the phases occurs across the APDs. At a voltage level around 300 kV the response may be ten times faster.

If there is a direct strike to the line there will be an exposed section of conductor at several hundred kV. Streamers will start to generate from the nearby non-struck phase – these will cause small holes to appear in the insulation as the electric field will be strongest next to the conductor – especially for the 50 mm^2 size. These streamers may be generated in several places up to 20 m from the struck point, but they will not as such cause conductor damage unless a breakdown occurs. It will take several micro-seconds for these streamers to cause a breakdown between the phases. A breakdown can occur here despite the large (500 mm) gap, because the lightning strike will have created a mass of ionised air at this point. This makes it easier for an arc to be established here than in the un-ionised air at the pole top.

An initial direct strike will cause some conductor damage, but it is unlikely to be terminal. However, unless the overvoltage surge can travel from the struck point to the nearest APDs and back again before mid-span breakdown, an arc will be established at the struck point. This arc will burn through the conductor in a matter of one or more seconds depending on the size of the follow through current.

The aim, therefore, is to have APDs that give a response rapid enough to reduce the return time of the reflected wave. In a low strike density area, it will be a matter of statistics as to the compromise between the chance of a direct strike to the line within the lifetime of the conductor and the frequency of APDs. The response time of the APD can be shortened by the use of short arc gaps as demonstrated by the design of the power arc device (PAD). Indeed, for phase–phase spacing greater than 700 mm at the pole, such arc gaps are not only preferred but are essential.

Finally, the response time of APDs can be delayed by impedance in the circuit e.g. at the contact point with the conductor and with too small a trace wire from the horn to the insulator.

18.13.7 Surge arresters

The problem with a surge arrester used to protect covered conductor lines is that it does not reflect the overvoltage surge as well as an operating arc discharge, but this effect is marginal in most cases. So the delay time before a mid-span strike point will receive the reflected wave back is mainly dependent on the distance travelled. The speed of the surge depends on various factors and may vary between 200 and 300 m/μs. In general terms, therefore, the wave will take between 0.6 and 1 μs to travel one span length and return. If the protection is every fourth pole then it is possible that the reflected wave will return in up to 2 μs for arrester protection but maybe double this or more for an APD system. The arrester may be connected to the line by some form of insulation piercing connector (IPC), but other suitable forms are the direct substitution of a line or tension insulator by a surge arrester or the current limiting arcing horn (CLAH) mounted under tension insulators (Figure 18.3). The latter unit will suffer some delays due to the presence of the arc gap.

Figure 18.3 The current limiting arcing horn (CLAH)

18.13.8 Frequency of protection

Lightning strike densities in the UK vary from 0.2 to over 2 str/km^2 yr and lightning currents from below 10 kA to several hundred kA. So the lightning protection has to either match the worst to be expected in area or be an economic compromise. Norwegian data indicate that if protective devices are installed every 200 m on covered conductor lines then a lightning strike to the line will cause phase–phase breakdown 50 per cent of the time.

The frequency of a lightning strike to the line and hence breakdown is dependent not only on the strike density but also on strike current. APDs, which depend on breakdown across two pin insulators, will be slow to operate and may not operate fully in some circumstances. This type of protection should be mounted more frequently to compensate for the delay in arc establishment by reducing the overvoltage wave travel time.

The general recommendation is that protection devices may be required on alternate poles. In most areas of the UK, pole-mounted equipment is present on average every seven spans. Surge arresters should protect this equipment and APD or PAD protection provided on alternate poles counting from the equipment pole . . . miss one, install one, miss one, install one etc.

18.14 Underground cables

One way to avoid lightning damage to overhead line conductors is to underground the system. But even this does not give 100 per cent protection and it has the disadvantage that lightning damage can often go undetected or at best is very difficult to locate and repair.

The previous section described how lightning is attracted to overhead lines. It can also be attracted to underground cables, generally leading to long-term damage but occasionally having catastrophic consequences. The ground normally behaves as a pure resistance at power frequencies, but at high frequencies it becomes a loose dielectric. At high impulse currents the resultant electric field can ionise the soil structure, producing areas of arcing and streamer formation underground. If the cable is within the streamer/arc area then the sheath can suffer multiple small punctures reducing its lifetime capabilities due to water ingress and corrosion.

The cable can also attract a direct strike in the same way as an overhead line if the distance, d, from the ground surface strike point is within the value:

$$d = k \cdot (pI)^{1/2} \tag{18.12}$$

where k varies from 0.05 to 0.08 depending on the soil resistivity, p (in Ωm) and with the current I in kA. A direct strike to the cable will result in the whole stroke current flowing through the metallic sheath of the cable. The underground cable will have a collection width of W metres given by:

$$W = (k^2 pI - r^2)^{1/2} \tag{18.13}$$

where r is the cable depth in metres.

Even if the cable does not suffer a direct strike, the soil can be turned into an electrolyte, resulting in the possibility of pinhole damage to the sheath and later water ingress. Overall, the damage to underground cables is difficult to quantify at this stage, due to the lack of basic research into the subject. However, the damage is likely to show itself not only by corrosion but also by leakage in oil-filled cables over many years, resulting in a reduced lifetime.

18.15 Sub-station protection

18.15.1 General

The protection system for any sub-station or transformer has the following aims:

1 to protect the equipment and its immediate area from a direct strike
2 to reduce the steepness and magnitude of the overvoltage surge before it reaches the equipment
3 to protect the terminations of the equipment with suitable devices
4 to eliminate any voltage enhancement due to reflected surges
5 to co-ordinate the overall protection package.

The equipment and nearby overhead lines can be protected against a direct strike by earthwires, poles and rods. However, on a tower line, if an earthwire is struck by lightning, then a back flashover across the phase insulators can cause an overvoltage surge to appear on the conductor.

The steepness of this overvoltage surge can be reduced by the application of protective devices at the penultimate pole (or more) before the equipment. This protection level must be co-ordinated with the protective devices at or within the equipment and at any open-end terminations.

18.15.2 Tower lines

The only SF_6 equipment liable to be associated with tower lines will be that contained in open or enclosed ground-based sub-stations. Sub-stations need protection from overvoltage surges arriving on the incoming or outgoing lines and also from direct strikes to the sub-station site. It is therefore necessary to protect the tower and wood pole lines in the region of a sub-station from direct strikes in order to reduce the overvoltage surges entering the station.

18.15.3 Reducing overvoltages

Overhead tower lines can be protected against direct lightning strikes by one or more earthwires (depending on the tower design) above the phase conductors. Earthing is critical. The earthwire must be earthed at each tower and, to avoid back flashovers across the insulators, the tower earth should be as low as possible. It is of critical importance that the final tower before a sub-station has the lowest possible earth impedance. This is best achieved by connecting to the sub-station earth. The lower impedance reduces the maximum voltage level induced by a lightning strike. Surge arresters at any OHL/cable junction are essential.

On wood pole lines with no earthwire, surge arresters at any OHL/cable junction must provide protection. For both tower and wood pole lines, care must be taken to avoid voltage doubling at any open point (e.g. open disconnector). As sub-stations are often different and also are commonly not constructed exactly as planned, individual examination of the lightning susceptibility and insulation co-ordination is recommended. If GIS equipment is involved then the cost of inspection can be minimal when compared to the cost of a failure.

18.15.4 Protective zone

In the case of a lightning strike the sections of line before and after an arrester will rise to voltages dependent on the initial voltage level, the length and type of line and the network characteristics (location of reflection points, equipment etc.). Essentially, a length of line close to the arrester will be protected, i.e. limited to voltages below its protection level. Further afield, the protection will be limited only to levels of double the arrester protection level. And this, of course, is in addition to any earth terminal voltage on the arrester. This concept of a protection zone around an arrester is therefore not a specific zone, but an area with a graded protection level determined by several factors. The area of protection up to the double level is determined by the rate-of-rise of the voltage wave, but the actual double level is independent of the voltage wave as long as the arrester is operational. The reflections from the arrester itself can affect the voltage level on the input side of the arrester. At some distance the voltage can rise to above the level of the original wave. This can lead to breakdown across an insulator or between phases that would not have occurred without the arrester present. This has implications for line protection at medium voltages when covered conductors are used and for pin insulator breakdown at poles near arrester protected equipment.

The actual protective zone of an arrester is dependent on:

- the number of overhead line feeders
- the local lightning strike density and current
- the lightning impulse protection level of the arrester
- the span length of the incoming line
- the acceptable failure rate for the station
- the overhead line outage rate for the first kilometre of line in front of the station.

The protective zone increases with:

- increasing difference between the withstand voltage and the protection level
- improved shielding to the overhead line
- reduced tower surge impedance.

The protective zone of a surge arrester can be calculated from the following:

$$L_p = V \cdot (0.91 \times U_{cw} - U_p)/2S$$

where U_{cw} is the BIL of system, kV, U_p is the lightning impulse protection level of the arrester, kV, S is the surge front steepness, kV/μs and V is the surge propagation velocity, m/μs.

It is assumed that in overhead lines $V = 300\,\text{m}/\mu\text{s}$ as the connections are presumed to be from bare wire OHL. In covered conductors the surge velocity may be up to 50 per cent lower than this. In cables V is assumed to be 150 m/μs.

The surge front steepness can exceed 3000 kV/μs at a point close to the strike. However, this value is attenuated and for co-ordination purposes a value of 1000 kV/μs is generally accepted. If a black flash over occurs close to the substation then the surge front steepness will be much higher, thereby reducing the size of the arrester protection zone.

Specific calculations for the number, position and design ratings of surge arresters to protect sub-stations can be made once knowledge of the above parameters is available. Details of cable lengths, the withstand levels of equipment within the station, open bus ends and any surge capacitances are also required. This allows the reflections with the station to be calculated according to travelling wave theory.

Normally, the entrance surge arresters should be sufficient to protect most stations at 132 kV and below, although some additional internal GIS arresters may be required. These arresters should have a higher protection level than the entrance arresters so that most of the input energy is absorbed at the station entrance.

In the case of transformers, overvoltage surges can enter directly or by transfer into the LV terminals. The voltage surge appearing at the HV terminals as a result of this surge is dependent on the waveform as well as the turns ratio and can be substantially higher than the initial voltage level. Surge arrester protection is therefore essential on all terminals of transformers associated with SF_6 insulated equipment. This is applicable to the use of SF_6 insulated vacuum switchgear especially as this type of equipment can cause some current chopping, resulting in high rate-of-rise waveforms appearing at the transformer terminals.

18.15.5 Insulation co-ordination

This chapter is concerned with protection from lightning strikes. It is possible that insulation co-ordination to protect against lightning overvoltages may be different to that employed as protection against switching overvoltages, depending on the equipment and the voltage level.

In the UK insulation co-ordination of SF_6 sub-stations up to 132 kV is often based solely on the use of arc gaps. This protection is only sufficient if the rate-of-rise of the voltage waveform is slow enough to allow the arc gap to operate before the SF_6 equipment insulation is damaged. Protection is, however, possible using co-ordinated surge arresters. The energy content of switching overvoltages actually determines the energy absorption capability of surge arresters at the upper end of the voltage range considered. At lower voltages this factor is less important compared with the heating effects of multiple lightning strikes. In the case of multiple strikes the rate-of-rise of the voltage is often higher for the second and subsequent strikes although the current level may be lower.

As has been described, the rate-of-rise of the voltage waveform depends on system configuration and surge impedance and on the current in the lightning strike. The distance of the strike from the equipment also has a significant effect as skin effects, and corrosion losses will reduce the steepness of the voltage rise. Typically, wavefronts can be slowed by 1 μs per kilometre of travel, although the exact figure depends on the conductor size and voltage/current levels.

It is therefore possible to use the known local lightning intensity in conjunction with network parameters to assess the likelihood of strikes to any SF_6 equipment location. In the case of 132 kV tower lines the local strike density and current level will determine the level of strikes to overhead lines within one kilometre of a sub-station and the possibility of shielding failure and also back-flashover. Network and distance considerations will determine the likely amplitude and steepness of the overvoltage surges. The frequency of these surges penetrating into the sub-station via the phase conductors can also be determined.

The connection between the overhead line and the sub-station (overhead or cable) and the branching network within the sub-station will then determine the position and value of surge arresters to protect the equipment economically against the average expected lightning storm.

It is critical to avoid thermal runaway problems with the arresters, and so the above data should also be used to determine the likely worst-case lightning strike over a set period. This period will depend on expected arrester life and sub-station cost. So the choice of arresters depends on the normal operational conditions (with switching overvoltages and temporary harmonics caused by line faults or load rejection), the overvoltage surges of the average expected lightning storm and the economic level of protection required against the worst case lightning scenario.

18.15.6 Surge impedances

The surge impedance of a sub-station can be substantially less than that of an overhead line, and the overall picture is also affected if an underground cable section is included. The cable is likely to have even lower surge impedances than the sub-station equipment and so high value reflection points are established. In lightning overvoltage situations, travelling wave effects in any cable sections between an arrester and connected equipment are extremely important. In general the voltage at the equipment will be higher than at the arrester. In order to increase the protection level, steps should be taken to reduce the rate-of-rise voltage and also to reduce separation distances between arresters and the equipment to be protected. If the cable is too short for significant attenuation of the wavefront rise time, then the effect of the cable connection will be to increase the vulnerability of the station by increasing voltage levels. It is possible therefore that surge arresters at the OHL/cable interface at the station entrance will not protect the sub-station. The cable/sub-station interface can have a voltage double the arrester protection level and interactions within the sub-station can increase this further. An underground cable connection can thus eliminate the sub-station from the protection zone of an arrester at the OHL/cable interface and leave it vulnerable to lightning damage.

It can therefore be necessary to use surge arresters at the station entrance and within the sub-station, especially for short cable lengths. If the cable is long enough to attenuate the voltage rate-of-rise and amplitude, then the OHL/cable interface arrester will be sufficient to protect the sub-station from incoming surges and no further protection would be required. This does not apply, of course, to direct strikes to the sub-station itself.

Effective shielding of the sub-station and nearby overhead line feeders will help to reduce the rate-of-rise of the voltage surge.

Although sub-station shielding will reduce the probability of steep-fronted waves within the station, there is a greater lightning catchment area for the overhead line feeders. Shielding these lines will reduce arrester currents, resulting in low residual voltages and improved protection.

18.15.7 Sub-station protection from a direct strike

18.15.7.1 General

The previous section covered sub-station protection from overvoltage surges entering via the overhead lines. Open sub-stations can also be protected from direct strikes. This can be done by overhead earthwires or lightning poles. In the UK protection from direct strikes is rarely necessary, but, in countries with a higher lightning incidence, such protection may be economically justifiable.

18.15.7.2 Overhead earthwires

All outdoor equipment (SF_6 switchgear, transformers etc.) can be protected from a direct strike by the use of overhead earthwires. The outer boundary of the protected zone is defined by an arc drawn from a centre, M, at a vertical height of $2H$ from the ground and $\surd 3H$ horizontally from the wire, where H is the height of the shield wire above ground level. The ground area protected by an overhead wire of height H is thus a rectangular zone of width $\surd 3H$ centred on the wire line. However, as the sub-station equipment is above ground level and in general not in a linear strip, a practical application is to use multiple earthwires. These have a protected area covering equipment above ground level between the wires. The distance between the wires can be calculated on the basis of the height of the line equipment. If two parallel earthwires of height H are separated horizontally by a distance C, then the protected zone between the wires will be defined by an arc drawn from a centre, height $2H$ and midway between the poles. The arc is drawn to pass through both earthwires. Outside the wires, the protected zone is as for a single wire. Typically, earthwires can protect line equipment 10 m above ground level 15 m high and up to 30 m apart.

18.15.7.3 Lightning poles

Similar protection can be obtained by using single poles instead of earthwires that may fall onto line equipment. The arc of protection (which is part of a conic section around the pole) is drawn from a centre $3H$ above ground (where H is the pole height) and $\surd 5H$ horizontally from the pole. The protected ground area is thus a circle of radius $\sqrt{5H}$ centred on the pole. If two poles are separated by a distance C, then the arc of protection is obtained from a centre height $3H$ and a distance $C/2$ from the poles horizontally. The surface of the protected zone is obviously a convoluted section that can be evaluated for specific situations. Typically, the greater angle of protection for poles (42°) compared to earthwires (30°) allows poles to be situated further apart, typically $3H$.

Broken shield wires can have serious consequences and shield wire maintenance requires an outage. In a recent survey, poles were quoted as being easier to install, more reliable and more aesthetic than shield wires. The poles used were generally metallic, although wood poles carrying lightning conductors could be a more economic and equally efficient alternative.

18.15.7.4 Arrester positions

In shielded sub-stations with unshielded overhead line feeders, the arrester should be mounted as close to the equipment as possible and preferably on the terminals. Multiple line feeders have the effect of reducing the steepness of overvoltage surges, but increasing their frequency of occurrence. In the course of a storm some incoming lines may be disconnected by faults. Flashovers to these lines can still damage sub-station equipment. However, it is recommended that the protection for multiple lines be at the same level as for single line feeders, i.e. line entrance arresters.

If the overhead line feeders are shielded then the voltages will be lower and less steep and it may be sufficient to protect the whole station by line entrance arresters alone.

18.16 Transformers

Pole-mounted transformers (PMTs) can suffer lightning damage in several ways. If there is a direct strike to the medium voltage line nearby then the overvoltage surge arriving at the PMT will have a sharply rising voltage on one phase. The voltage may be rising at several hundred kV/μs. This is likely to pass the arc gap before it has a chance to operate and therefore the surge will enter the HV winding directly. Two scenarios are now possible. At this effective high frequency (several MHz) the impedance of the copper winding is very high (due to skin effects) and so the voltage will be attenuated rapidly. Hence there will be substantial voltage differences between the adjacent turns in the outer winding and this can lead to arcing between the turns (Figure 18.4), which are normally only insulated to 1 kV. There will also be a high impulse current through the winding, generating substantial EMF forces trying to send the inner core into orbit. As the core is part of the frame, it is somewhat reluctant to be ejected and so instead the HV winding is shunted out of position (Figure 18.5). In both these scenarios the damage is instant and fatal. The high voltage can also cause internal arcing and the follow-through current can cause burnout of the HV leads.

The damage described above can also occur with a nearby indirect strike when high voltages are induced on all three phases. In this case, the voltage rise will be less steep and the arc gaps may operate before the BIL of the equipment is reached.

Figure 18.4 Inter-turn arc damage

Figure 18.5 Shunted PMT winding

However, the gap will discharge to the tank and the current will flow to the tank earth. It can be seen therefore that the LV insulator is now the lowest BIL point and a flashover here is likely. This will cause a voltage surge on the LV windings that will be transformed up and several hundred kV could be present on the inner turns of the HV windings. This situation can result in similar arc damage between the turns as is the case for a direct strike. Follow-through current will also erode the arc gap electrodes at around 1 cc/kA s. So a fault current of, say, 1 kA – which is extremely common – will erode the length of a 6 mm diameter electrode at around 35 mm a second.

A neutral point arrester can be fitted between the neutral point and the tank body. The idea is to limit the internal voltage stresses so that there is no flashover to the LV terminal and therefore no surge on the LV winding. Under the values of tank voltage measured on an 11 kV transformer with an LV BIL of 30 kV the protection has been demonstrated to work successfully even with arc gap protection on the HV terminals. The arrester should fail open-circuit. On a 20/22 kV system (as used in north east England and Ireland) a higher rated arrester would be required. Any practical arrester with this capability would limit an 8/20 μs 5 kA lightning current to a 36 kV peak. A 20/22 kV transformer should thus have an LV impulse withstand of at least this value.

The use of these arresters can be justified on a technical and economic basis as only one small arrester is needed to provide protection.

18.17 Summary

Lightning is generated from local movement of warm air, often associated with weather fronts and local topography. It can be in a negative or positive form, negatives strikes being of short duration with low current levels and generally in the summer months, whereas positive strikes can be high-current long-duration events that often occur in the winter. An efficient lightning protection strategy depends on a suitable overall protection policy based on the local conditions and the characteristics of the network. The choice and positioning of the actual protection devices (arresters, triggered arc gaps, arc gaps) can be vital in allowing protection to do its job. Covered conductors have to be treated as a piece of equipment that needs protection as lightning can present particular problems not present on bare wire lines. Protection for line and equipment should also be considered within a co-ordinated policy that includes quality of supply considerations. The use of neutral point arresters for PMTs is a recognised economic lightning protection system.

18.18 References

1 COORAY, V.: ‘The lightning flash’ (IEE Publishing, 2003)
2 ‘Thunderstorms’, *Weather*, 1996, **51**(12), pp. 371–376
3 CEGB report RD/L/R980, 1961
4 MIYAKE, K. *et al.*: ‘Characteristics of winter lightning in Japan’, *IEEE Trans. Power Deliv.*, 1992, **I**(3), pp. 450–456
5 EATON, D.: EA Technology Ltd Lightning Detection System, archived data

6 RUSCK, S.: 'Induced lightning overvoltages on power transmission lines with special reference to the over-voltage protection of low voltage networks', *Trans. Royal Inst. Tech.*, 1958, **120**
7 ANDERSON, R. B., and ERIKSSON, A. J.: 'Lightning parameters for engineering applications', *Electra*, March 1980, **69**, pp. 65–102
8 MOUSA, A. M., and SRIVASTRAVA, K. D.: 'The implications of the electrogeometric model regarding effect of height of structure on median amplitude of collected lightning strokes', *IEEE Trans. Power Deliv.*, April 1989, **4**(2), pp. 1982–1989
9 IEEE Working Group: 'Estimating lightning performance of transmission lines', *IEEE Trans. Power Deliv.*, July 1993, **8**(3), pp. 1254–1267
10 GOLDE, R. H.: 'Lightning' (Academic Press, London, 1977) vol 2, chap. 22
11 ERIKSSON, A. J.: 'The incidence of lightning strikes to power lines', *IEEE Trans. Power Deliv.*, July 1987, **2**(3), pp. 859–870

Chapter 19

The future

19.1 Introduction

19.1.1 General

This chapter looks at the future trends in overhead line design, construction and maintenance and also at draft regulations and how these may affect the current situation. The future is something that can be anticipated and precautions taken – current data can be extrapolated and the present situation gauged in terms of overhead line supply quality and economic performance. For the future, the challenges of climate change and new regulations and practices require preparations now. The future trend for overhead lines has to be based on two areas – what sort of environment the overhead line is likely to face for the next 50 years and how supply quality can be maintained costeffectively.

19.1.2 Scope

This chapter will therefore look at four areas:

1. new products
2. climate change and how it affects overhead line design in the UK
3. the present protection armoury, before considering the new techniques becoming available
4. finally, the chapter will see how these new developments can best be put to use to provide a total environmental protection design package.

19.2 New products for the operation of future overhead distribution networks

19.2.1 Composite materials

The introduction of composite materials on to overhead distribution networks can be traced back to the early 1960s when the first composite insulators were introduced.

These insulators offered many advantages over existing porcelain and glass systems. Some of these included a greater resistance to damage from vandal attack, lighter weight and most significantly an improvement in performance in polluted environments. Unfortunately, these early insulators were significantly more expensive and were often found to have poor long-term field performance. In contrast, today's composite distribution insulators are significantly lower in cost (often competitive with porcelain and glass) and have demonstrated good long-term service performance. At the bottom end of the medium voltage range (11–20 kV) these insulators are just becoming competitive in price in comparison with porcelain or glass pin/post insulators, and composite tension insulators are now in standard use. The antivandal advantage of polymeric material can make it even more costeffective in reducing fault levels. At 33 kV and above, composites are already capable of providing good performance at competitive prices.

In addition to insulation applications, there are now a number of structural composite alternatives available for use within modern distribution networks. These include lightweight alternatives to heavy steel cross-arms (up to 75 per cent lighter) and for distribution poles. Composite poles and cross-arms can bring more reliable long-term performance, increased strength and reduced weight. One big problem with poles is that they are difficult for linesmen to climb as spikes cannot be used. However, modern safety regulations on climbing on wood poles may lead to suitable methods for climbing composite poles. An advantage is that they can be precisely defined in terms of strength, whereas wood poles have a wide range of mechanical properties. However, the high cost means that they will only be used for niche applications at the present time.

Future developments could combine composite insulation and structural systems to form more simplified line support systems. In addition, further developments in composite materials will enable lower cost systems to be produced which retain the same performance benefits over the conventional materials used in today's distribution networks.

19.2.2 Robotic and electronic tools

Investigations into the potential uses for robotic devices on overhead networks have revealed three main areas. These are:

1 to aid and enhance overhead line inspection
2 to assist in overhead live line working operations
3 for tree cutting.

In addition to robotic devices, recent developments in information technology have provided a basis to establish a range of new electronic tools aimed at improving the efficiency and effectiveness of maintenance operations. Already in other areas of the distribution network these tools have been used to carry out and provide expert knowledge and assistance for non-expert electricity company staff to service and maintain electrical plant. As well as offering expert advice, these devices can also be used to collect significantly more information on the condition of plant and equipment as maintenance is carried out. This could provide the operators of overhead distribution networks with a greater level of confidence in determining the likely future performance of the network.

Future overhead networks could also benefit from the development of devices aimed at determining the condition and physical performance of overhead lines in the field. A number of such devices have already been developed and their use is increasing on the UK distribution network. One example of this is the ultra-sonic system and the drilling and computer evaluation (see chapter 12) for assessing the degree of rot present in wood poles and the extent of residual strength left. In the laboratory there are now a number of other bench scale devices aimed at determining the performance of electrical insulation which could be adapted for field use to once again provide fast and accurate condition assessment of overhead lines in the field.

19.2.3 Conductors

One problem facing the industry is to match current conductors in service with the work load generated by population movements and changes in industry patterns. Increasing power transfer by allowing conductors to rise in temperature has its limits both metallurgically and in terms of regulatory ground clearance.

In order to increase the reliability while maintaining adequate performance of conductor systems used within overhead distribution networks, there are now many developments available. The new conductors becoming available for transmission systems that could possibly be used in future distribution networks have been summarised in chapter 9.

There are also currently three separate material developments being investigated to enhance the physical performance of aluminium conductors. These are:

1. the development of new aluminium alloys
2. the development of surface treatments for corrosion resistant aluminium conductors
3. the development of bimetallic conductor systems.

Research and development into new alloys continues to maximise physical properties and has yielded a wide range of options leading to increased performance in one or more factors required for conductor systems, i.e. strength, conductivity or corrosion resistance. The introduction of new permanent surface treatments for aluminium conductors will enable the use of stronger and more electrically efficient alloys which do not need to be optimised for corrosion resistance. Some preliminary work has already demonstrated on a laboratory scale that significant improvements can be made to a conductor's corrosion resistance by the application of a low-cost surface coating during manufacture.

For a number of years bimetallic aluminium and copper conductors have been available. These conductors are made up from individual bimetallic wires consisting of a central high tensile steel core surrounded by either aluminium or copper. Manufacturers claim that the overall tensile strength of these conductors can be increased significantly over standard conductor systems without compromising electrical conductivity.

19.3 Applications for modern composite materials within overhead distribution networks

19.3.1 Background

A composite material comprises two or more materials whose individual properties are combined together to produce a desired set of properties. Over the last 50 years, the increasing use of plastics has led to the introduction of a wide range of polymer composite materials.

Today the use of polymer composites has increased such that they now account for the vast majority of man-made composite materials and as such are often simply referred to as composites. These composites were initially formed by combining the properties of polymers with inorganic fillers. The nature of the individual polymer and inorganic materials obviously had a marked effect on the overall properties generated in the composite. Some of the first polymer composites developed were based upon elastomer or rubber compounds. This has led to the development of materials for many important modern day applications; for example, the rubber compounds used to manufacture tyres.

The introduction of composites based upon thermosetting polymers has led to a vast increase in their use in structural applications. Today thermoset resin-based composites have replaced more traditional materials such as metals and ceramics because they can be specifically tailored to the needs of an application. In the majority of cases these composites are used because:

- they are lighter in weight
- their mechanical performance can be specifically tailored to suit an application
- they can be moulded into complex shapes
- they offer a lower cost alternative to more traditional materials.

In addition to simply replacing the properties of more traditional materials in applications, composites can and often do provide additional properties that can lead to greater design flexibility and an improved performance within an application. In many cases this results in more simplified design options requiring fewer separate components that can often lead to a more reliable performance at a lower cost of manufacture. The most striking examples of this are seen in the boat building industry where single moulded composite hulls are now commonplace. Another example of this approach is seen in the automotive industry where designers have simplified the design of cars by replacing steel bumpers and some areas of the steel bodywork with large single section composites.

19.3.2 Composite insulators

Some of the earliest polymer composite insulators became available in the mid 1960s offering benefits such as reduced weight and a wider range of design options compared with traditional glass and porcelain insulators. In the early days of composite insulator technology many problems were encountered in achieving the desired mechanical and electrical performance coupled with their resistance to environmental degradation. Since then many manufacturers have succeeded in developing composite insulators capable of delivering good performance over many years. Moreover, in high pollution areas composite insulators have been shown to outperform insulators manufactured from porcelain and glass.

The main reasons for the use of composites world-wide are their performance under polluted conditions, the fact that they are less attractive to vandalism (composite insulators do not shatter when shot) and their low cost. Silicon rubber was by far the most popular choice, with 90 per cent of insulators of this material and only five per cent of EPDM. Europe and Asia are the main silicon users, with Australia only choosing silicon for 68 per cent of its applications (23 per cent EPDM).

Overall, silicon rubber-based composite insulators have given good performance at all voltages although there are specific areas where they are not suitable (e.g. near fertiliser factories). The performance of these insulators depends not just on the silicon but also on the type of filler used and this is dependent on the individual manufacturer. Experience in the field has indicated that lifetimes of 30 years are currently being achieved when appropriate choice of insulator material and specific creepage distance has been made. Modern composite insulator materials are now far superior to those that have been in service for several decades and greater lifetimes can thus be expected. Lightweight and cheap composite insulators for distribution networks are commonly based on polyethylene (e.g. HDPE) and are comparable in price to porcelain.

One of the driving forces behind the continued improvement in performance of composite insulators over the years has been the development of a wide range of experimental techniques to determine their likely full lifetime performance. Often also referred to as accelerated ageing, these techniques involve the identification of the most significant factors in the environment that the insulators would be expected to encounter. The effect of increasing the level of exposure of composite insulators to each of these factors is then studied in an attempt to determine the likely performance over a full service life. As the performance of composite insulators in the field has improved over the years, so this has led to improved accuracy in accelerated ageing by creating benchmarks for accelerated performance. Today there are a wide range of standards used to categorise the performance of composite insulators under laboratory accelerated ageing or high pollution conditions.

19.3.3 Structural use of composite materials in overhead distribution networks

From the very early days there have been three main functional components used in structures for overhead distribution networks, i.e. electrical insulators, cross-arms and support poles. As outlined above, in the mid 1960s composite insulators were introduced as alternatives to porcelain and glass.

In recent years composite materials have been used to develop alternatives for cross-arms and support poles. Manufacturers claim polymer composite structures can be economically erected in remote regions using helicopters due to their high strength to weight ratio, and this has been demonstrated by utilities in the USA. They also claim that polymer composite structures offer benefits over conventional wood/metal structures by having:

- a lower overall cost
- a lower maintenance demand (e.g. they are not susceptible to rot problems)
- a lighter weight
- more flexibility in service
- better dielectric properties
- greater fire resistance
- increased structural strength
- a more environmentally sound alternative to treated wood poles.

Composite poles and cross-arms have been evaluated for their structural integrity both in the laboratory and in exposed field locations. In this way the electricity utilities and manufacturers have gained an insight into the suitability and likely long-term performance that these composite systems can offer in a range of distribution applications.

19.3.4 Optimised designs for composite insulating distribution structures

As outlined previously, in other industries the full benefits of the properties that can be engineered into composite materials are often realised only when a different design approach is adopted. These designs often make use of the fact that many complex component parts within designs using traditional materials can be replaced in a simplified design using fewer composite components. In all cases this leads to lower manufacturing costs, easier installation and a more reliable performance due to the reduction of interfaces between the various components.

19.3.5 Optimised composite insulating materials for low-cost insulators and insulating distribution structures

The most critical area for the continued electrical performance of an outdoor composite insulation system is its surface. Using this rationale, there are developments in new surface technologies initially to produce lower cost insulators, with the potential of impacting later on the design of optimised composite distribution structures.

In the first development, surface modification can be achieved by developing a silicone coating material that will be chemically bonded to a low-cost composite insulator substrate as a stage in the manufacturing process. It is envisaged that this approach will enhance the surface properties of relatively cheap polymer composite substrates (e.g. filled epoxy or polymer concrete) such that outdoor insulating components could be produced of equivalent or better overall performance to the currently more costly components manufactured from single component bulk materials such as silicone or EPDM elastomers.

The second development utilises recent advances in fluoropolymer synthesis technology to achieve a cross-linked highly fluorinated polymer surface. It is felt that this novel approach has the potential to deliver ultimate surface performance for polymer composite insulator systems. It is believed that a fluorinated insulator surface could be achieved such as with a bulk PTFE material without the problems previously associated with the processing of PTFE insulators. This could be achieved by doping the uncured filled polymer composite material with a modified fluoro-polymer prior to moulding the insulator. As the insulator is cured, the fluoro-polymer would then migrate to the surface to yield the desired surface properties, thus affecting the mechanical properties of the composite substrate.

The perceived benefits from these developments are:

1 Reduced material costs enabling the manufacture of composite distribution insulators at costs.
2 Potentially higher performance than composite insulators currently available.
3 Greater design flexibility, e.g. could be used to give track resistant properties only where required in composite insulating support structures, i.e. potentially removing the need for conventional distribution insulators.
4 Can be used to provide the surface properties of conventional bulk silicone insulation materials to lower cost composite alternatives.

19.4 Materials selection

19.4.1 Introduction

Wood pole overhead distribution networks have been in existence for over 120 years. Today's networks still have much in common with the materials, construction and operation used in the first networks. In recent years advances in materials technology have made available a much wider range of material choices which can be used to construct modern overhead distribution networks. The selection of the most appropriate materials for specific components within overhead distribution networks is largely dictated by their cost against the delivery of a closely defined performance. The following aims to identify the required performance and the range of materials available for each of the three main components of an overhead distribution network (i.e. the insulator, the pole and the cross-arm). In each case the materials used most commonly at present will be identified along with the pros and cons of the other materials available.

19.4.2 Poles and cross-arms

19.4.2.1 Traditional construction

Wood poles and steel cross-arms were used in some of the first overhead distribution networks and are still widely used today. Almost all line construction standards are based upon the properties of these materials. This has led to these standards dealing almost exclusively with the relative size and shape requirements of wood poles and steel cross-arms to deliver a specific mechanical performance.

In the UK, treated Scots Pine is generally specified for wood pole distribution applications. This is chemically treated before installation to reduce the effects of rotting (see chapter 15). Wood poles are graded into four broad strength classifications for a range of heights from 6 to 24 m in length. These classifications (light, medium, stout and extra stout) are broadly based upon calibrated ratios of pole diameter to pole height, although light poles are no longer used at medium voltages. Unfortunately, because of the nature of wood poles (i.e. they are a naturally occurring product) there is a significant degree of natural variation in the mechanical strengths of wood poles within these broad classifications. This problem is then further compounded through the service life of the pole as the effects of natural rotting affect wood poles to varying degrees. In order to counter this, the mechanical performance ranges relating to the wood pole classifications indicated above are generally a factor of four greater than the actual mechanical performance requirement for most line constructions.

Typically heavy angle section galvanised steel is used for cross-arm applications within the UK overhead distribution network. As BS 1990 defines wood pole selection, ENATS 43-95 specifies the size/design and nature of steelwork used to manufacture cross-arms.

19.4.2.2 Composite alternatives

In addition to simply replacing the properties of more traditional materials in structural applications, composites can and often do provide additional properties that can lead to greater design flexibility and an improved performance within an application. In many cases, this results in more simplified design options requiring fewer separate components, which can often lead to a more reliable performance at a lower cost of manufacture.

In recent years composite materials have been used to develop alternatives for cross-arms and support roles. Modular light weight polymer composite constructions can be economically erected in remote regions using helicopters due to their high strength to weight ratio. The extra cost of the pole is off-set by the cheaper installation. Manufacturers claim polymer composite structures offer benefits over conventional wood/metal structures by having:

- a lower overall cost
- a lower maintenance demand (e.g. they are not susceptible to rot problems)
- a lighter weight
- more flexibility in service
- better dielectric properties
- greater fire resistance
- increased structural strength
- a more environmentally sound alternative to treated wood poles.

Typically, composite poles are hollow and either tapered or parallel (dependent upon the method of manufacture used). The composite materials used are commonly glass fibre reinforced polyester resins with wall thickness varying from 5 to 15 mm dependent upon the pole's required mechanical performance. Composite cross-arms are generally manufactured as hollow box section profiles with wall thickness of around 5 mm. These composite structural alternatives are generally available at a higher cost compared with the traditional wood and steel. Unlike wood poles, however, composite poles are manufactured to a specific mechanical performance characteristic that can be accurately maintained throughout the lifetime of the pole in service. This more predictable mechanical performance compared with wood poles could allow composite poles to be used which are closer to the actual in-service mechanical requirements.

19.5 Line design for global warming

19.5.1 General

Overhead lines (OHL) are by their very nature open to everything the environment can throw at them (see chapter 16). From ground level to the conductors the system has to withstand mechanical and biological forces that cause many headaches to OHL designers and supply customers alike – not to mention the emergency repair teams that bear the initial brunt of any major catastrophe.

Ground conditions affect foundation capability with corrosion of towers and fungal growth and rot in wood poles. Wind loads affect the pole and conductors as well as any inconveniently planted trees that tend to fall onto or grow into the lines. Wind blown debris causes electrical shorts as does conductor clashing. And that is before the 'real' weather such as lightning, freezing rain, ice and wet snow attacks.

Environmental protection for overhead lines will become even more important in this climatically changing world. Several areas of protection policy from the past are either out of date already or will soon be, and as climate change takes effect so the OHL engineer needs to be aware of present climatic trends to maintain reliability. The evaluation of climatic parameters is a key factor for the design, upgrading, reinforcement, operation and maintenance of the power supply network. These climatic trends indicate ten per cent increased wind and ice loads and 15 per cent increased lightning activity over the next 50 years. However, protection needs to be focused to be truly effective and be subject to a cost–benefit analysis. Protection can be improved by upgrading outdated policies and by using new models that are becoming available to check out the existing network and new plans in the light of the current and future climate.

19.5.2 Line design in the past

The most important issue for the overhead line (OHL) design engineer is to ensure that lines now being constructed, re-built or refurbished are capable of providing the level of reliability expected by the customer and the regulator. It was necessary in the past only to consider the design standards and historical performance of existing overhead lines and, in particular, the mechanism of failure. It was then possible to design the line economically to reduce the probability of failure to an acceptable level.

The initial design parameters introduced in 1896 were substantial but 'a bit on the expensive side'. Standards in the UK swung totally the other way in 1942 when no allowance was made at all for radial ice on small conductors. In 1946 the infamous BS 1320 standard was introduced and this, in various guises, forms the basis of most 11 kV OHL construction in the UK; cheap but effective in most situations. However, in some cases, mechanical overload of the conductors led to high fault rates and the improvements of ENATS 43-10 and 43-20 (for heavy duty 33 kV lines) were introduced. Finally, a major change came in 1988 with the publication of ENATS 43-40. Revision of this specification has been necessary in the light of the publication of CENELEC standard BS EN 50423. The future for UK line design is therefore tied in now with European developments.

19.5.3 Environmental protection

There is a need to focus protection systems and to apply the correct protection in the correct place, i.e. protection that matches the expected fault conditions whether these be due to ice, snow, wind, lightning or whatever, in the precise environmental (e.g. local pollution) and electrical (e.g. earthing, position or structure) conditions. An appreciation of past history and exactly what is required is essential for an all round economic protection package. In particular, answers to certain questions are required:

- Why is protection needed in this area?
- How important is it to protect this area (number or importance of customer, frequency of bad weather)?
- What type of protection would be most suitable (antigalloping devices, covered conductors etc.)?
- Why did previous protection fail?
- What would be the consequences of failure?
- What can be done to limit the effect of failure?
- What is the cost of protection relative to the cost of failure?
- Is there a protection problem or a network problem?

A blanket policy of having just one type of protection system or design to cover all eventualities can be uneconomic and may not answer the problems.

19.6 Wind

19.6.1 Wind loads on conductors

The effect of wind on conductors is in several forms:

- wind pressure
- vibration
- galloping.

Wind loading alone rarely causes line failure, although this can happen if the wind loads are not understood properly and the line design is then inadequate. Wind loads vary according to conductor type (stranded, compacted, bundled, covered) whereas most design packages use only a pure conductor area.

Conductor failure is not common. There may be, on average, 2000 conductor failures in the UK each year through storms or corrosion, out of a network of over 600 000 conductor kilometres. The main cause of HD copper conductor failure is normally overloading by snow/ice, tree problems, clashing (leading to arcing) in high winds. Copper is corrosion resistant and easily repaired, so long lifetimes can be expected. It also has a high vibration tension limit. The major problem with many small copper lines (16 mm^2) is lack of capacity leading to necessary re-conductoring.

Aluminium conductors (AAC, AAAC and ACSR) are subject to corrosion problems of the aluminium and loss of zinc on any steel core used along with the additional problem of lower vibration tension limits, leading to fatigue failure. In many DNOs, however, the main cause of conductor failure is weakening due to crevice or galvanic corrosion.

19.6.2 Trees blown down onto conductors

Tree contact with overhead lines is always a bad thing and can be a contentious area. Lightning voltages can jump from the line to a tree in close proximity or touching, and trees can fall onto or grow into lines or cause faults by intermittent contacts in windy conditions. It is difficult to see how avoiding tackling this problem head on with a good quality tree-cutting programme can do anything but reduce reliability. The use of XLPE covered (but not fully insulated) conductors can improve reliability by allowing the supply to stay on in many cases until the trees are removed. The use of fully insulated conductors such as LV ABC and the Swedish medium voltage aerial cables can change tree-cutting policies and improve reliability while not offending landowners.

19.6.3 Clashing

XLPE covered conductors can eliminate the problem of clashing and some tree problems but have an increased susceptibility to vibrational fatigue and lightning. Recent tests have shown that corrosion can be avoided and suitable operational and design considerations can lead to reduced vibration and lightning problems. Covered conductors can use AAAC, AACSR or even copper as their conducting core.

19.6.4 Wind on ice

Wind loading on bare and iced conductors has been included in design calculations for many years. However, there are many variations in how wind on an ice envelope can affect the conductor loading. Recent work has shown reduced conductor tension levels under blizzard conditions when different conductor support devices are used. There are over 12 years of accumulated data at UK test sites on a variety of conductors from tiny telecommunication cables to EHV conductors of over 40 mm diameter, as well as covered, twisted and sheath modified systems. This database should provide both wind only and wind on ice loads for nearly 40 different conductors.

19.6.5 If it's not one thing . . .

Antivibration devices either as additional features or built into the conductor can prove very successful. However, such devices can alter the snow/ice loads and also the general wind loading. Recently, a vibration control device worked extremely well at no extra cost but led to increased wind loads and frequent galloping – even at high tensions.

19.7 Snow/ice

19.7.1 ENATS 43-40 Issue 1

Chapter 5 has covered the UK's historical approach to snow/ice loads. Basically, the UK was split into various weather zones where mechanical loads on the overhead line were based on a wind (factor 1 to 6) and ice (factor A to E) loading. These loadings were based on historical weather data for the past 50 or more years. However, it can be misleading to rely totally on historical data and ignore climate change.

19.7.2 Weather mapping

The UK has been split into different weather loading areas and altitudes (England/Wales above/below 300 m and Scotland above/below 200 m). Each zone has its own wind and ice loading which is applied to the overall line design package to work out the maximum conductor pressure (MCP), maximum conductor tension (MCT) and the maximum resultant force on the conductor (MCR). These are generally based on non-aerated ice, but other icing models can be considered according to BS EN 50423-3-9. Currently, European Union projects such as COST 727 aim to produce improved wind/ice maps and predictive now-casting of severe weather that could affect overhead lines.

19.7.3 Is it snow or ice?

Wet snow causes the UK supply industry most problems. Freezing rain is also a major source of poor reliability in some areas. However, the most common form of weather load is rime ice. This forms readily on lines at any height but is most common at land heights above 100 m. The accretion envelope is different from that for wet snow, often being more aerodynamic and hence suffering less wind loading. BS EN 50423-3-9 is the first UK adopted standard that recognises that ice loads are not necessarily the density of an ice cube (917 kg/m^3) and that wet snow generally has a density of 850 kg/m^3 and rime ice around 510 kg/m^3. This may lead to a new approach to snow/ice load evaluation in the future.

19.8 Reliability for the future

19.8.1 General

The acknowledgement of global warming is almost universal. It may be that the current UK regulations view the 'circumstances in which they are used' as applying to the weather likely to be experienced in the future on the basis of current knowledge. Any line that is refurbished, re-built or newly constructed may not be considered 'fit for purpose' if the OHL engineer keeps his head in the past and does not acknowledge what the future may hold.

19.8.2 Global warming trends

There is considerable evidence now [1] that there will be a global warming of between 1.5 and 4.5 °C over the next 50 years. In the UK the changes should result in:

- a temperature increase of 2 °C
- an increase in absolute humidity of five to ten per cent
- increased cloudiness
- increased precipitation
- cooler stratosphere
- warmer, more humid surface (leading to increased convective potential energy)
- no change in solar heating
- increased convective instability.

These will occur over the next 50 years. So, what effect will they have on the weather elements that cause OHL faults?

19.8.3 Precipitation

Increased cloudiness and humidity will lead to increased precipitation. This can occur as an increase in general rainfall but is also quite likely to occur through more frequent convective storms [2]. A potential fingerprint of global warming is the observation that extreme rainfall events are increasing [3]. These heavy downpours are leading to major flooding events world-wide and have occurred in the UK within the last few years.

19.8.4 Ice

Rime ice occurs in inland and upland areas and can be severe over 100 m above mean sea level in the UK. The increased temperature of the lower atmosphere implies that the air is able to carry more water vapour. The amount of rime ice is dependent on the water vapour content of the atmosphere and so ice loads can be expected to increase by five to ten per cent. It is not clear how this will affect glaze ice events at the moment.

19.8.5 Wet snow

Snow levels depend on the absolute (as opposed to relative) humidity. In the formation of snow flakes, water droplets or minute ice crystals depend on nuclei to form. These nuclei may be particles only a few microns across. In the UK the presence of sea salt nuclei in the atmosphere is critical to snow formation. Increased storm activity can increase the number of these nuclei and the higher absolute humidity means that more snow crystals can form. Wet snow will still fall in the same temperature range, i.e. +0.5 to +2 °C, but the increased source material may increase snow levels by up to ten per cent.

The increase in the number and severity of isolated convective storms can also increase wet snow blizzards. Single cell convective storms are not a problem as they rarely last more than 30 minutes in one place. Multi-cell storms, however, have the ability to renew themselves by creating new storm cells just behind the parent storm. This has the effect that one area can receive continuous heavy local precipitation lasting several hours [4]. It is not possible at this time to put a figure on the increase in snow loads likely from these convective storms. However, the evidence is strong that localised severe blizzard conditions will occur more frequently over the next few decades.

19.8.6 Lightning

It is accepted that lightning frequencies world-wide increase with temperature since the deep convection thunderstorm systems are directly linked to the transfer of momentum, heat and moisture in the atmosphere. Model and field data [5] indicate that a climate with twice the present day concentration of CO_2 will have a 30 per cent global increase in lightning levels. An analysis of European data [6] has indicated specific trends in lightning activity. The northern latitudes, including the UK and northern Germany, have already shown increases in lightning activity relative to France, Switzerland, Austria and southern Germany. The UK may thus start to experience the lightning storms presently typical of northern France.

19.9 What can be done?

19.9.1 Probabilistic design

The semi-probabilistic line design points the way forward. This system uses probabilistic meteorological information as the input to link lightning effects and snow/ice loads for specific areas. That link can be upgraded to allow for the developing trends in the climate over the next few decades. This will allow a risk of failure to be balanced against the cost of refurbishment/construction for new and existing lines. The way forward is two-fold:

1 Improvement in the basis for the evaluation of expected lightning levels and snow/ice loads and development of models for lightning prediction and wet snow accretion on overhead lines.

2 Application of probabilistic methods for mechanical and electrical design incorporating risk analysis and risk management.

In addition, work has concentrated over the last few years on producing a new lightning model based on the physics of the lightning strike and the probability of its strength, frequency and location. This includes the differences in conductor type and configuration as well as drop-in modules covering spurs, inset cables, stay wires, soil resistivities, arc gap and arrester protection, earthing and pole-mounted equipment etc. – indeed all the features of a typical section of network. As the predicted lightning changes are fed into the model, the performance of the network and likely fault levels under various improvement scenarios can be evaluated.

19.9.2 Line design models

The use of the wet snow and lightning models described earlier can be combined with the existing wind loadings to produce an overall design package for overhead lines based on probabilistic principles. This can be used to check out existing lines to see how 'fit for purpose' they are for the area and weather conditions predicted, as well as to plan new lines with specific risk levels.

19.10 Action

19.10.1 What action can be taken now?

The aim of this section is to try to improve reliability by reducing the uncertainty of the database for the design and maintenance of overhead lines. The OHL engineer needs to:

- introduce climatic adapted design
- produce more efficient and cheaper operation and maintenance
- improve reliability.

This can be achieved by using existing models and:

- evaluating the performance of the present system using higher snow/ice, wind and lightning levels
- modifying existing lines where necessary on a risk management basis
- designing new lines to meet the expected climatic changes.

19.10.2 Action plan

The specific action plan should:

1 use design models now available with ten per cent increased wind and ice loads and 15 per cent higher lightning intensities
2 be aware of new models being developed to improve estimation of lightning levels and wet snow loads from meteorological data
3 monitor developing models for climatic change as they affect overhead line reliability.

19.11 Summary of preparations needed for climate change

1 Lightning protection can be improved by upgrading protection policies presently based on out-dated information and equipment.
2 Protection needs to be focused to be truly effective and be subject to a cost–benefit analysis.
3 The evaluation of climatic parameters is a key factor for the design, upgrading, reinforcement, operation and maintenance of the power supply network.
4 The OHL engineer needs to be aware of present climatic trends to maintain reliability.
5 These trends indicate ten per cent increased wind and ice loads and 15 per cent increased lightning activity over the next 50 years.
6 Models are available to enable the OHL engineer to check out the existing network and new plans in the light of the current and future climate.

19.12 Risk

In the future, risk evaluation may be the norm. This does not mean being risky, but a realistic risk assessment can also be costeffective. The tendency for many recent standards has been to provide ways of estimating risk on which to base future strategies such as:

1 Problems
   - deterioration with time
   - reliability loss with time.
2 Strategies
   - condition-based strategy
   - risk-based strategy
   - basis for decisions.

Chapters 11–13 and 15 have defined various coping strategies for optimising future policies and providing a basis for decision making. The need for new strategies comes about because:

- environmental considerations mean that new OHL will not be built to any great extent
- so, more is needed out of existing OHL
- some lines are operating at their thermal or capacity limit
- in the next few years many lines installed in high growth periods will be reaching the end of their theoretical lives.

The actions that are possible are:

1 structural capability
   - design strength
   - loads

   - reliability
   - maintenance level.
2 electrical capability
   - voltage level
   - current carrying capacity
   - outage frequency
   - outage duration.
3 upgrade
   - increase structural reliability
   - decrease failure probability.
4 uprate
   - larger conductors
   - compact conductors
   - higher conductivity conductors
   - lower sag conductors.

Various issues to contend with could be:

1 deterioration due to
   - type of conductor
   - quality of conductor
   - internal protection
   - external environment.
2 obsolescence due to
   - lack of (or too many) spare parts
   - lack of skilled linesmen.

The options include a condition-based strategy:

- condition of component
- its importance relative to the network
- replace according to condition/importance
- prioritise
- information.

And a risk-based strategy with the following criteria for action:

- investment costs
- investment period
- maintenance/operational costs
- benefits: operational and safety improvements
- risk.

This strategy can include risk reduction by:

1 action before failure:
   - uprate
   - refurbishment (restore design life)
   - refurbishment (extend design life)
   - increase maintenance.

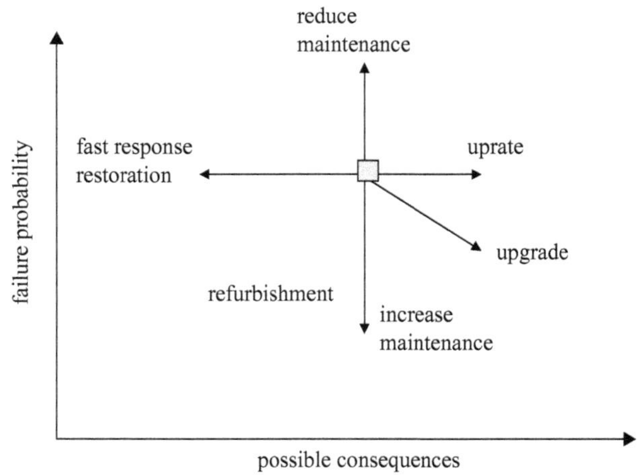

Figure 19.1 Risk strategies

2 action after failure:
   - fast restoration (bypass or generation)
   - fast response/repair.

or a risk acceptance strategy based on:

- no change to maintenance
- reduce maintenance
- do nothing at all.

All these will have an effect on the failure rate of the network and may have other consequences, particularly financial. Figure 19.1 looks to encompass this dilemma in a single graph.

19.13 References

1 Intergovernmental Panel on Climate Change (Cambridge University Press, New York, 1990)
2 CRANE, R. G., and HEWSON, B. C.: *Int. J. Climatol.*, 1998, **18**, pp. 65–76
3 KARL, T. R. *et al.*: *Bull. Am. Meterol. Soc.*, 1996, **77**, pp. 279–292
4 'Thunderstorms', *Weather*, 1996, **51**(12), pp. 371–376
5 YARNAL, B. *et al.*: *Weather*, 1999, **54**(1), pp. 19–25
6 PRICE, C., and RIND, D. J.: *Geophys. Res.*, 1994, **99**(DS), pp. 10823–10831

19.14 Further reading

MACKERRAS, D. J.: *Geophys. Res.*, 1985, **90**(D4), pp. 6195–6201

Glossary

AAAC	all aluminium alloy conductor
AAC	all aluminium conductor
AACSR	aluminium alloy conductor steel reinforced
ABS	air break switch
ACAR	aluminium conductor aluminium reinforced
ACD	anticlimbing device
ACSR	aluminium conductor steel reinforced
Al	aluminium
APD	arc protection device
AR	auto recloser
BIL	basic impulse level
BM	benchmark
CAD	computer-aided design
CC	covered conductor
CENELEC	European Committee for Electrotechnical Standardisation
CLAH	current limiting arcing horn
Cu	copper
DNO	distribution network operator
EDS	everyday stress
EDT	everyday tension
EMF	electro-magnetic field
ENATS	Energy Networks Association Technical Specification
EOL	end of life
EPR	earth potential rise
ESI	Electricity Supply Industry
ESQCR	Electricity Supply Quality and Continuity Regulations
FL	field liner
FOS	factor of safety
GMR	ground mounted recloser
GPS	global positioning system
GSP	global satellite position

HD	hard drawn
HDPE	high density polyethylene
HI	health index
HSE	Health and Safety Executive
HV	high voltage
HVCC	high voltage covered conductor
IAD	insulated access (aerial) device
INS	inertial navigation system
IPC	insulation piercing connector
LA	local authority
LEMP	lightning electromagnetic pulse
LPS	lightning protection system
LV	low voltage
MCP	maximum conductor pressure
MCT	maximum conductor tension
MCW	maximum conductor weight
NGT	National Grid Transo
NNA	National Normative Annexe
NRPB	National Radiological Protection Board
OCB	oil circuit breaker
OHL	overhead line
OPGW	optical pipe ground wire
PAD	power arc device
PC	personal computer
PE	polyethylene
PME	pole-mounted equipment
PMS	pole-mounted switchgear
PMT	pole-mounted transformer
PPE	personal protective equipment
ROEP	rise of earth potential
ROW	right of way
RSM	rolling sphere method
RSV	residual strength value
SEF	sensitive earth fault
SSSI	site of special scientific interest
SVD	spiral vibration damper
TAG	triggered arc gap
TP	transient pulses
TSG	triggered spark gap
UGC	underground cable
UTS	ultimate tensile strength
XLPE	cross-linked polyethylene

Index

www.ingramcontent.com/pod-product-compliance
Ingram Content Group UK Ltd.
Pitfield, Milton Keynes, MK11 3LW, UK
UKHW020235290926
15470UKWH00002B/568